MICROELECTRONIC DEVICES

MICROELECTRONIC DEVICES

2nd Edition

Keith Leaver

*Imperial College of Science,
Technology and Medicine*

Imperial College Press

ICP

Published by

Imperial College Press
516 Sherfield Building
Imperial College
London SW7 2AZ

Distributed by

World Scientific Publishing Co. Pte. Ltd.
P O Box 128, Farrer Road, Singapore 912805
USA office: Suite 1B, 1060 Main Street, River Edge, NJ 07661
UK office: 57 Shelton Street, Covent Garden, London WC2H 9HE

British Library Cataloguing-in-Publication Data
A catalogue record for this book is available from the British Library.

Cover design: Courtesy of Mr Peter Cheung

First edition was published by Longman Group UK Limited in 1989.

MICROELECTRONIC DEVICES (2nd Edition)

ISBN 1-86094-013-7
ISBN 1-86094-020-X (pbk)

Printed in the United Kingdom.

Contents

Preface to First Edition

In an age in which the electronic engineer becomes more and more concerned with system design and performance, the time allotted for the study of the physical mechanisms of operation of electronic devices grows ever smaller. The aim of this book is to provide what the author regards as the barest minimum needed for an engineer to understand future developments in fabrication and design. It is introductory rather than exhaustive, and deliberately brief, in order that the student may grasp essentials without becoming swamped by either excessive detail or lengthy explanations.

It covers the physical mechanism of conduction in semiconductors, using models which are made quantitative, and describes the operation of the three most significant devices used in integrated circuits: the *p–n* junction, the bipolar transistor and the MOSFET. Devices such as varactor diodes, LEDs and photodiodes are mentioned but not covered in detail, as they are readily grasped if only the three basic devices, together with the physical properties of semiconductors, are properly understood.

Students who have studied some electronics in an 'A' level physics course will probably be familiar with two models of a semiconductor: the valence bond model — in which the electrons are treated as either bound in 'bonds' or free — and the energy model — in which the distinction is made on grounds of electron energy. The first task is to use these models to make some quantitative predictions of behaviour.

We begin with the valence bond model, as it is simple and provides sufficient understanding for the discussion of free carrier motion, so that the factors controlling the resistivity of a semiconductor can be treated quantitatively.

We then develop the energy band model in Chapter 2, in more detail than at 'A' level. This model is more useful for discussing concentrations of carriers and the action of junctions between semiconductors. This model has varying levels of sophistication, but this book will concentrate on the most elementary of these.

The need for a model with which to make quantitative predictions is, of course, basic to both science and engineering. Later in the book we shall construct circuit models of transistors which can be used for designing complete amplifiers or switching circuits. The semiconductor models we begin with have been devised by physicists to fit as closely as possible all aspects of observed behaviour, and we shall only explain enough for the student to use them rather than to understand completely their justification. For the latter, more advanced books on solid state physics must be consulted.

It is a pleasure to acknowledge the support of my many colleagues who have contributed to the teaching of this subject at Imperial College, and without whom this book would have been the poorer. Among them I wish particularly to thank Dr P. M. Gundry, whose pioneering of the teaching of semiconductor theory has strongly influenced my own presentation. I should also like to thank Mrs. W. Ibsen for much help with the typing.

Keith Leaver
July 1988

Preface to Second Edition

I am grateful to Imperial College Press for the opportunity to revise this book, which has been popular as an introductory book with students within the Department in which I teach.

Although the later chapters have undergone substantial revision, the intention remains of providing an introduction to the subject which concentrates primarily on imparting understanding of physical concepts. Thus, while the ideas underlying computer modelling are covered, this book does not attempt to teach the full details of computer models for the devices discussed here. The justification remains the same as previously: I believe that a good grasp of the contents will enable any competent engineer to use more advanced texts with ease, and even to begin using commercial modelling packages with confidence.

I would like to thank various colleagues anew: Trevor Thornton for encouragement and many discussions, and John Cozens and Colin Vickery for assistance in clarifying the text. I also wish to thank Richard Lim of IC Press for his faith in the book, and patience with my revisions.

Keith Leaver
Imperial College, London
May 1996

Chapter 1

Conduction in Semiconductors

1.1 Conduction by electrons and holes: the valence bond model

It is almost a commonplace that the outer (valence) electrons in a metal are free to move through the regular lattice of atoms which make up a single (perfect) crystal of a metal. The contrast with conduction in an electrolyte, e.g. salt solution, is that no matter is transported when the current flows. In a metal we intuitively expect the current, and hence the conductivity (i.e. 1/resistivity) to be proportional to the number of electrons available for carrying current.

We can model a metal (Fig. 1.1) as a close-packed arrangement (a lattice) of spherical ion cores, between which is a 'gas' of electrons which can move freely in response to the electric field set up inside it when a voltage is applied across the solid.

By contrast, a very pure piece of silicon (or other semiconductor) has a conductivity at room temperature of the order of 10^{-4} S/m, or 10^{12} times smaller than that of copper. The valence electrons in Si are thus unable to carry a current when an electric field is applied. (We shall see later that a *very* small number of them can, so explaining the finite conductivity.)

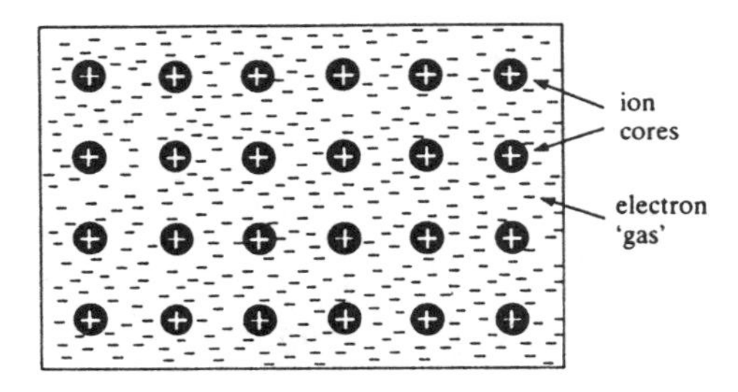

Fig. 1.1 A metal can be modelled as a close-packed lattice of spherical cores held together by the attraction of the 'gas' of valence electrons.

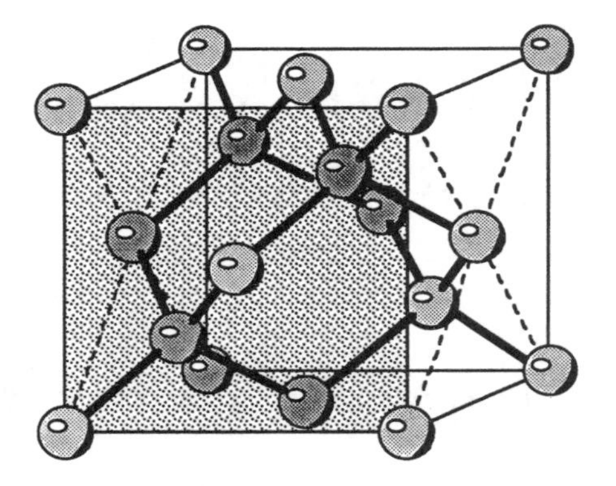

Fig. 1.2 A semiconductor lattice is held together by covalent bonds each containing two electrons. In Si and Ge there are four bonds per atom.

Our model of a Group 14/IV semiconductor (C, Ge, Si) is that of a lattice of atoms held together by covalent bonds as in Fig. 1.2. (Chemists divide the bonds between atoms into several classes, of which the covalent is the strongest, and has rigid directions.) Each bond between two Si atoms is made from an electron donated by each of the two adjacent atoms. Clearly the electrons paired up in these covalent bonds are unable to conduct an electric current. It is as if these electrons were unable to move, although this view is strictly incorrect, as we shall see later.

The conductivity of a semiconductor can be controllably increased by 'doping' the pure material with a small proportion of a suitable impurity. For example, the solid line in Fig. 1.3 shows how the conductivity σ of so-called *n*-type silicon rises almost in proportion to the concentration N_D of arsenic, phosphorus or antimony atoms per m^3 across six orders of magnitude. Since these elements are in Group 15/V of the periodic table they each possess one more valence electron than silicon or germanium, which are in Group 14/IV. Figure 1.3 strongly suggests that each Group 15/V atom contributes one electron which is free to conduct electricity. The simplified lattice diagram of doped silicon in Fig. 1.4 shows how a Group 15/V impurity forms four covalent bonds to neighbouring Si atoms, leaving its fifth valence electron only weakly bound by electrostatic attraction to the single excess

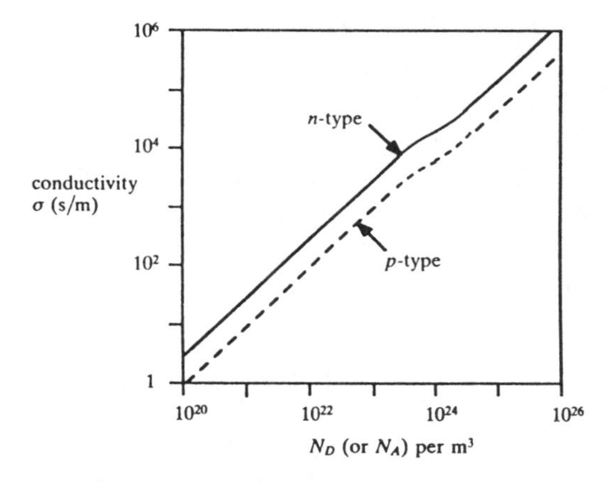

Fig. 1.3 Conductivity of silicon plotted on log–log scales against concentration of As, P or Sb (solid line) or B, Al; (dashed line).

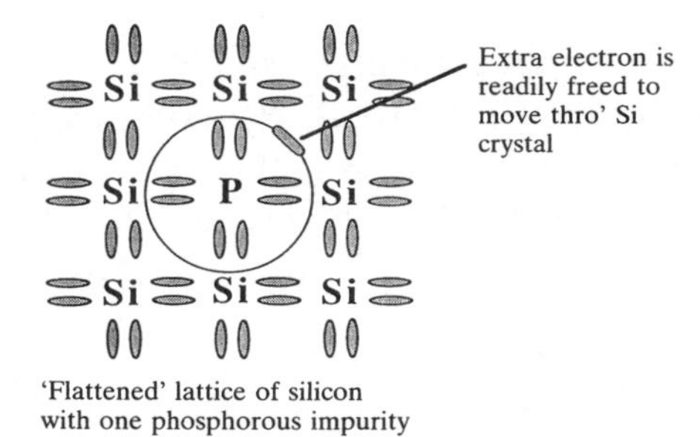

Fig. 1.4 Illustrating how a 5-valent impurity donates a 'free' electron to lattice atoms.

positive charge on the nucleus. As this fifth electron is rather easily freed from its parent nucleus, we can roughly explain the conductivity plot in Fig. 1.3, if we assume that the current flow and hence the conductivity are directly proportional to the concentration of free electrons.*

The extra electron is readily freed, because the force attracting it to its parent As^+ ion at a distance r, as given by Coulomb's law, is $e^2/4\pi\varepsilon_r\varepsilon_0 r^2$ and is reduced by the permittivity ε_r of the lattice through which the electron moves. In silicon, $\varepsilon_r = 11.7$, and the small force is readily overcome by the thermal, vibrational, energy always present in a solid at room temperature — something we shall frequently need to take account of.

Terminology: We have described above what is called an n-TYPE semiconductor, in which As, P, Sb are called DONOR impurities. It is sometimes called an EXTRINSIC semiconductor, to distinguish it from the pure, or INTRINSIC semiconductor, which contains by contrast a very low concentration of free electrons. That concentration is finite only because the thermal energy of vibration of the atoms frees a very small proportion of electrons from the bonds between them.

We use the symbol n for the concentration (i.e. the number per unit volume) of electrons in a semiconductor, and the symbol N_D for the concentration of donor impurities. Hence:

$$n = N_D \text{ in an } n\text{-type semiconductor.}$$

This is true provided that N_D is not too large ($\lesssim 10^{24}$ m^{-3} in Si) or too small (greater than the concentration of electrons in the intrinsic semiconductor).

1.2 *p*-type semiconductors

If impurities such as boron or aluminium from Group 13/III of the periodic table are substituted for the Group 15/V impurities, the plot of conductivity vs concentration is also nearly linear, as shown by the dashed line in Fig. 1.3. It is a little lower than, but nearly parallel to, the solid line representing the n-type semiconductor. We call the Group 13/III doped material a p-TYPE semiconductor.

---

*Note that, at this point, we are turning a blind eye to the deviations from linearity in Fig. 1.3: the first aim in constructing the model is to get the gross features correct, and to leave the detail for later attention.

Using a simple lattice model as before, Fig. 1.5(a) shows how the Group 13/III atom has one fewer electron than is needed to fill the four bonds with neighbours. Thus a deficiency in one bond appears to aid conduction of electricity about as much as the donation of an electron by a Group 15/V atom.

We can understand this if we assume it is rather easy for an electron from a neighbouring bond to fill the original half-vacant bond as in Fig. 1.5(b), leaving a 'hole' in the silicon bonding network. This 'hole' appears to move

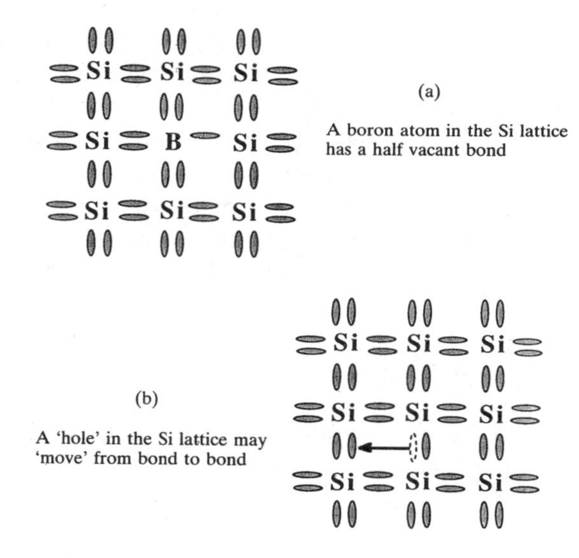

Fig. 1.5 The formation of holes in silicon: (a) Group 13/III atom bonded into lattice (b) creation of hole by filling half-empty bond on impurity atom.

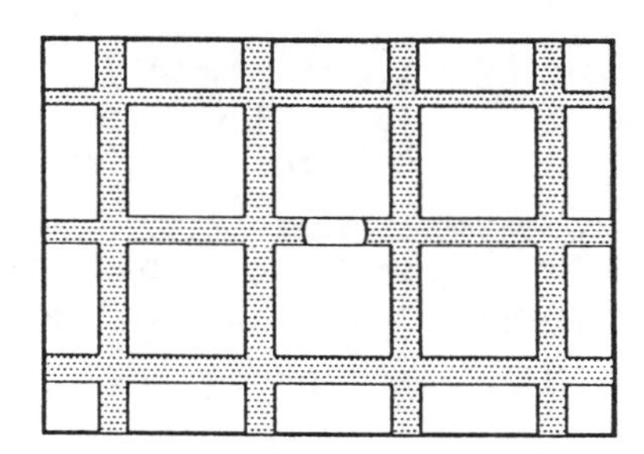

Fig. 1.6 A bubble analogy for the motion of holes.

yet further if neighbouring electrons jump across in sequence to fill the empty slot. Note that motion of the hole to the right in the diagram results in one more electron ending up on the left, and is equivalent to the carriage of one positive charge, equal to that of an electron, along with the hole towards the right.* The motion of the hole can most easily be appreciated by analogy with that of an air bubble (Fig. 1.6) in a lattice of interconnected pipes (the 'bonds') containing water (the valence electrons). When full, the flow of water in the pipes to right and left must just balance — i.e. no net flow occurs, as in an intrinsic semiconductor. A bubble (hole) is necessary for flow to occur.

Just as a small attractive force held the fifth electron to its donor atom unless freed by thermal energy, so the same applies here. The free hole with its local positive charge experiences a weak attractive force F toward any Group 15/V atom which has accepted an excess negative electron charge in creating a free hole. Being small, this force is normally unable to overcome the thermal agitation present at room temperature, and the hole remains free to 'conduct' electricity. Thus a p-type semiconductor, as it is called, is made by doping with ACCEPTOR impurities, and a concentration N_A of acceptors gives rise to an equal concentration of holes, the symbol for which is usually p, thus

$$p = N_A \text{ in a } p\text{-type semiconductor}$$

(This is true provided that N_A is within the same limits as given for N_D in an n-type semiconductor, i.e. between about 10^{17} and 10^{24} m^{-3} in Si.) The conductivity of the solid is proportional to N_A, and hence to p, suggesting that each hole contributes to the flow of current. When we come to study the Hall effect in Chapter 2, we shall be able to show that each hole does indeed carry a positive charge, and that it can be thought of as a 'particle' which is free to move in a similar way to a free electron.

1.3 Intrinsic semiconductors and minority carriers

An intrinsic semiconductor, not being quite an insulator, contains a small concentration of free electrons at room temperature. Since this concentration

*Note that, in the immediate neighbourhood of the hole, the lattice actually contains an excess positive charge, being a neutral lattice from which an electron has been removed.

has been created by freeing electrons from bonds, an equal number of holes must also be present — i.e. $n = p$. We shall use the symbols n_i and p_i to represent these values in an intrinsic semiconductor, so that

$$n_i = p_i \text{ (INTRINSIC)} \tag{1.1}$$

In silicon, $n_i = p_i = 1.4 \times 10^{16}\ \text{m}^{-3}$ at 300 K, and the value rises very strongly with temperature, which 'shakes' more electrons free. Values of n_i for other semiconductors at 300 K are given in Table 1.1.

Table 1.1

	Ge	Si	GaAs
n_i (m^{-3})	2.5×10^{19}	1.4×10^{16}	9×10^{12}

Since thermal vibration is not unique to intrinsic materials, there must be a small number of holes present in n-type materials, and a few free electrons in p-type materials. The possibility of both carriers co-existing is vital to the operation of devices, and we shall need to find ways of calculating the concentrations of both the MAJORITY carriers and MINORITY carriers, as they are called.

Note: MAJORITY carriers are *electrons* in n-type, holes in p-type semiconductors.
MINORITY carriers are *holes* in n-type, electrons in p-type semiconductors.

For the present we take their numbers for granted, and discuss how to derive an expression for the conductivity.

1.4 Motion of free electrons in semiconductors

It is easy to comprehend how the uniform motion of a 'cloud' of free electrons leads to the flow of current. But the motion in practice is more complicated, for, even in the absence of a current, the free electrons possess kinetic energy by virtue of the thermal energy in the solid. They behave like

PANEL 1.1

The kinetic theory of gases

The introductory treatment found in many school textbooks shows that the pressure of a gas can be derived from the rate of change of momentum of molecules as they bounce elastically against the walls of the container. The resulting equation for the pressure P of a gas having n molecules per unit volume, each of mass m, is

$$P = \frac{1}{3} nm\overline{c^2}$$

where $\overline{c^2}$ is the mean square velocity of all the molecules, calculated irrespective of their direction of motion.

Multiplying P by the volume V of the one mol of gas, we find

$$PV = \frac{1}{3} N_0 m\overline{c^2}$$

where $N_0 = nV$ is the total number of molecules in a mol of gas (Avogadro's number).

Comparing with the universal gas equation $PV = RT$ leads to the conclusion that

$$\frac{1}{3} m\overline{c^2} = \frac{RT}{N_0}$$

from which eqn. (1.2) follows directly by putting $k = R/N_0$

a gas, with random velocities which can be calculated using the kinetic theory of gases (see Panel 1.1).

One of the results of kinetic theory is that the average kinetic energy of each atom of mass m is directly proportional to the absolute temperature, according to the equation:

$$\frac{1}{2} m\overline{c^2} = \frac{3}{2} kT \tag{1.2}$$

where k is Boltzmann's constant, value 1.38×10^{-23} J/K, and $\overline{c^2}$ the mean square thermal velocity of an atom.

By analogy, the electron gas should also obey eqn. (1.2), so that we can use it to calculate the root mean square velocity of an electron in terms of its mass m_e:

$$\sqrt{\left(\overline{c^2}\right)} = \sqrt{\left(\frac{3kT}{m_e}\right)} = 1.2 \times 10^5 \ \text{m/s at } T = 300 \text{ K}$$

This value is approximately correct for a semiconductor, but is quite incorrect for any metal.*

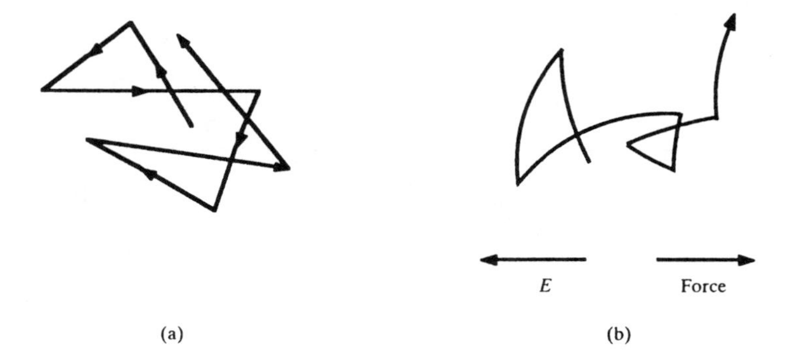

Fig. 1.7 Illustrating the thermal motion of an electron (a) no electric field (b) with electric field towards the left.

As in any gas, an electron's velocity is random in both direction and magnitude when no voltage is applied, and is as frequently positive as negative when measured with respect to a fixed direction. So the *average* velocity of an electron, observed over a long enough period, is zero. Likewise the instantaneous average velocity of all the electrons is zero. If we follow

*Free electrons are so close together in a metal that they interact strongly and do not behave like a classical gas. Their quantum properties make their average energy very much greater than $\frac{3}{2}kT$. Hence their velocities are much higher than in a semiconductor. Thermal energy still makes an additional contribution to their average energy, of value approximately kT.

the motion of an individual electron, its velocity changes magnitude and direction whenever it collides with an atom, as illustrated schematically in Fig. 1.7(a) (collisions with other free electrons are much less frequent as they are so few in number). The path is straight between collisions, unless an electric field is applied when the acceleration it causes makes the paths curved, as shown in Fig. 1.7(b) (what shape has each path now?).

Since the acceleration is always in the same direction, the electron acquires an average velocity along this direction, and this, called the DRIFT VELOCITY, is proportional to the current flowing. The reason that this velocity does not increase continuously is that each collision, being random in its effect, causes all memory of the direction of the drift velocity to be lost. Conservation of energy implies that the random velocity increases slightly (i.e. the solid gets hotter), but the increase is negligibly small for normal currents.

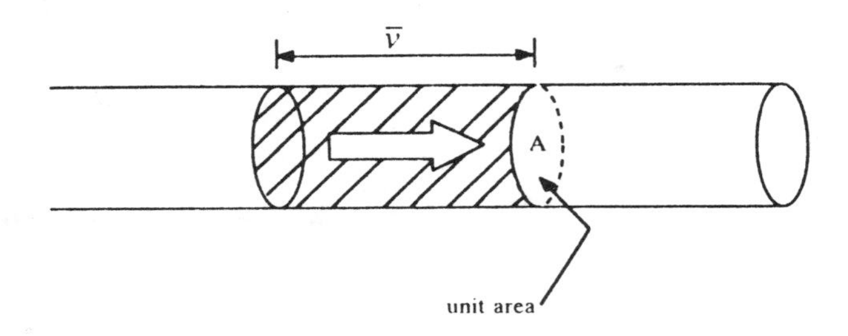

Fig. 1.8 Showing the calculation of current due to electrons flowing with drift velocity $\bar{v}$.

To relate the current to drift velocity, assume that a cloud of n electrons per m^3, each with charge $-e$, is moving at a uniform drift velocity $\bar{v}$ down a wire of unit cross-sectional area (Fig. 1.8). The current equals the charge carried by all the electrons crossing plane A in one second i.e. the charge on all electrons upstream of and within a distance $\bar{v}$ of it. The number of electrons in the shaded volume is $n\bar{v}$, and the charge is thus $-ne\bar{v}$. Hence the current per unit area — called CURRENT DENSITY — is

$$J = -ne\bar{v} \tag{1.3}$$

(i.e. J is opposite in direction to $\bar{v}$)

From this equation it is easy to find that $\bar{v}$ is typically about 10^3 m/s in a piece of Si in which $n = 10^{23}$ m^{-3}, when it carries the rather large current density of 1000 A/cm^2.

We see that the drift velocity is normally much smaller than the random thermal velocity calculated earlier, and so the random thermal motion is barely altered by it. Hence, we shall attempt to calculate $\bar{v}$ in terms of the collisions of electrons, assuming that the frequency of collision is unaltered by the drift velocity.

At each collision an electron loses the average drift velocity it had during free flight, and has zero drift velocity immediately after collision. Hence the change in the average drift velocity is $\bar{v}$ per collision. If the electron spends a mean time τ between collisions, it collides $1/\tau$ times per second, losing an amount $\bar{v}/\tau$ of drift velocity per second, i.e.

$$\left(\frac{dv}{dt}\right)_{\text{collisions}} = -\frac{\bar{v}}{\tau}$$

This loss just balances the gain in velocity between collisions given by Newton's second law*:

$$\left(\frac{dv}{dt}\right)_{\text{field}} = -\frac{eE}{m_e}$$

As the overall change in v is zero in equilibrium, we can equate the sum of these rates of change to zero, giving on rearrangement:

$$\bar{v} = -\frac{e\tau E}{m_e}$$

The magnitude of the drift velocity in unit field is called the MOBILITY, symbol μ_e, so that

$$\mu_e = \frac{e\tau}{m_e} \tag{1.4}$$

*Since the electron is not moving in empty space, its acceleration is slightly modified by the shielding effects of the charges around it. We could allow for this by giving the electron an effective mass, different from its free mass m_e, but we ignore this correction here for simplicity.

μ_e is a ratio of velocity to electric field strength, and so is measured in units of (m/s) per (V/m), or m^2/Vs.

The current density due to the electrons is thus

$$J_e = -ne\bar{v} = ne\mu_e E$$

The conductivity σ is just J/E (see Panel 1.2) so that, assuming negligible numbers of holes:

$$\sigma = \frac{J_e}{E} = ne\mu_e \tag{1.5}$$

We can use this equation to estimate μ_e for silicon using the graph in Fig. 1.3. For example $\sigma = 2$ S/m when $N_D = 10^{20}$ per m^3. Then

$$\mu_e = \frac{\sigma}{ne} = \frac{\sigma}{N_D e} = 0.12 \text{ m}^2/\text{Vs}$$

A more accurate measurement at low doping gives the result

$$\mu_e = 0.135 \text{ m}^2/\text{Vs} \text{ (Silicon)}$$

Values for other semiconductors are given later, in Table 1.2.

Terminology: a summary

- $\sqrt{\left(\overline{c^2}\right)}$ The THERMAL VELOCITY of an electron is that random component due to thermal agitation present in all bodies.
- $\bar{v}$ The DRIFT VELOCITY of an ensemble of electrons is their mean velocity in response to an electric field.
- μ_e The MOBILITY is equal to the drift velocity in unit field.
- τ The MEAN FREE TIME is the mean time between collisions of an electron with the lattice of atoms.
- λ The MEAN FREE PATH is the mean distance travelled between collisions.
- m_e The MASS of an electron.

PANEL 1.2

Conductivity and current density

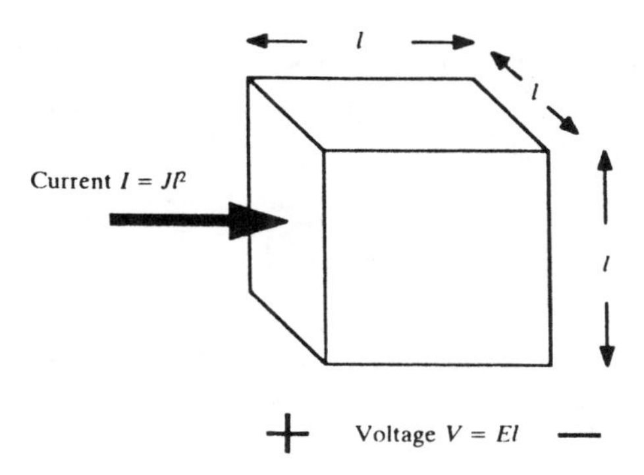

The figure shows a cube of material side l, through which flows a current density J, normal to one face.

The current is $I = Jl^2$, and the voltage between opposite faces is $V = El$, where E is the electric field strength in the cube.

The resistance R of the cube is expressible in terms of resistivity $\rho = 1/\sigma$:

$$R = \frac{\rho l}{l^2} = \frac{1}{\sigma l}$$

Hence

$$\sigma = \frac{1}{lR} = \frac{1}{l}\left(\frac{I}{V}\right)$$

Using the expressions above for I and V gives

$$\sigma = \frac{1}{l}\left(\frac{Jl^2}{El}\right) = \frac{J}{E}$$

1.5 Factors determining mobility and conductivity

It might be thought that an electron would collide with almost every atom in turn, since the atoms are closely packed in a solid, at a mean separation of 0.27 nm in silicon.

We can see that this is not so by estimating the mean distance between collisions, often called the MEAN FREE PATH, λ. It is related to τ simply through the thermal velocity*

$$\lambda = \tau\sqrt{\left(\overline{c^2}\right)} \tag{1.6}$$

In silicon, whose electron mobility is 0.135 m^2/Vs we find using eqns. (1.4) and (1.6) that

$$\tau = 0.8 \text{ ps} \quad \text{and} \quad \lambda = 90 \text{ nm}$$

Thus λ is much larger than the interatomic separation.

It can be shown that free electrons collide only with *defects* in the regular lattice of atoms in a crystal. A perfectly regular lattice allows electrons to pass freely because of their wavelike nature (their 'quantum' properties), just as a light wave travels unhindered through a crystal of rock salt, but is strongly scattered by table salt. Thus the mean free path between collisions is a measure of the distance between defects or distortions of the atomic arrangement.

These are of two kinds: those which are permanent, primarily impurity atoms (e.g. ionised donors, but also displaced Si atoms), and those which are transient, i.e. atoms which are momentarily out of place because of thermal vibration (see Panel 1.3).

In contact with a gas at temperature T, it is not surprising that the atoms vibrate with an average kinetic energy equal to that of the gas atoms, i.e. $3kT/2$.

The total number of collisions per second is thus the sum of the frequency of collisions with impurities $1/\tau_I$ and the frequency $1/\tau_L$ of collisions with lattice vibrations:

---

*The ways that velocities and free times are spread among the electrons differ. Thus the product of their averages, $\tau\sqrt{\left(\overline{c^2}\right)}$ is not the same as the average of their products, which equals λ. Thus eqn. (1.6) is only approximately correct.

$$\text{i.e.}\quad \frac{1}{\tau} = \frac{1}{\tau_I} + \frac{1}{\tau_L} \tag{1.7}$$

In a pure, carefully prepared single crystal of Si or Ge, permanently displaced atoms are rare and $\tau = \tau_L$. However, in a material such as 'polysilicon' (used as a conductor in integrated circuits) which is composed of many small, randomly oriented grains or crystallites and is termed POLYCRYSTALLINE, electrons colide at the crystallite boundaries where the atomic arrangement is irregular. Moreover, because of the way polysilicon is made, each tiny crystallite is far from perfect internally. So the mean free time is much smaller than τ_L and is determined by the concentration of defects, which control τ_I. So the mobility of electrons in heavily doped polysilicon is low compared with that in single crystal Si, and is typically 0.03–0.04 $\mu m^2/Vs$.

PANEL 1.3

Phonons: sound waves fuelled by thermal energy

Atoms in solids cannot move without stretching or compressing the bonds they make with their neighbours. The figure shows a simple model in which the bonds are represented as springs, the atoms as masses, forming a sort of 'atomic mattress'. It is clear that the vibration of one atom causes a vibrational wave motion to propagate through the solid. Hence thermal energy is present in the form of 'sound' waves moving randomly in the lattice of atoms. Their characteristic frequencies are high — about 10^{13} Hz — and these waves are sometimes called PHONONS.

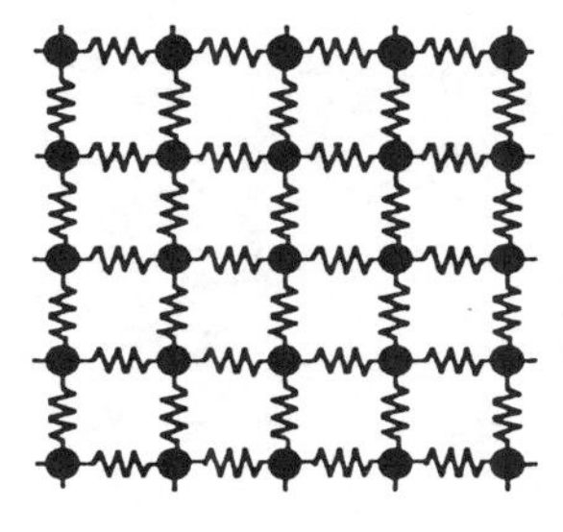

As an example of an intermediate case, heavily doped silicon is interesting, because we can partly explain the deviation of Fig. 1.3 from linearity. At high donor concentrations ($\geq 10^{23}\,m^{-3}$) collisions with ionized donor (or acceptor) impurities are frequent, but not much more frequent than collisions with lattice vibrations. The mobility is thus reduced below the 0.135 m^2/Vs usually quoted for silicon at a temperature of 300 K, and is approximately halved at a doping of about $3 \times 10^{23}\ m^{-3}$.

At the high doping densities used in emitters of bipolar transistors and the *p–n* junctions of GaAs LEDs and lasers, these effects are even more noticeable.

1.6 Temperature dependence of τ_L and μ_e

Because τ_L and τ_I depend on temperature in different ways, it is possible to identify their relative contributions to τ. Measurements show that in weakly doped germanium the mobility of electrons varies almost as $T^{-3/2}$, while in silicon the relation approximately obeys a $T^{-5/2}$ law. The graph in Fig. 1.9 shows the behaviour as a function of both temperature and doping.

We can obtain a qualitative appreciation of the temperature dependence of τ_L from a rather simple model of the vibrations of an atom. If we assume

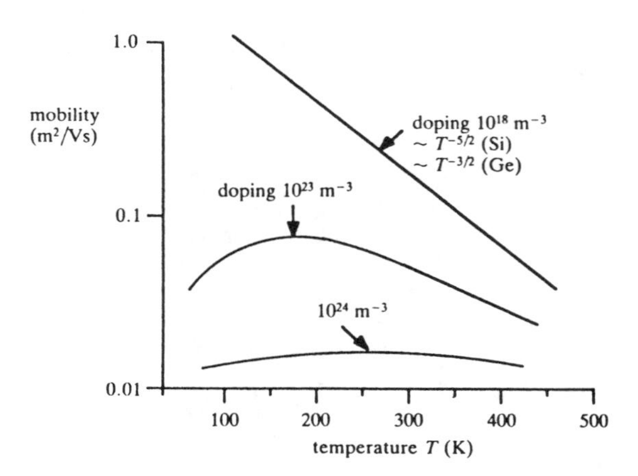

Fig. 1.9 Approximate behaviour of mobilities in Si and Ge with temperature.

that it oscillates in random directions but at a constant amplitude, a, which depends on T, then the mean free path between collisions may be assumed inversely proportional to the cross-sectional area which this sphere presents to the oncoming electron, i.e. to a^2 (see Fig. 1.10). But a^2 is likely to rise proportionately to temperature, according to the following argument.

The kinetic energy of a mass undergoing simple harmonic motion is proportional to the square of the amplitude of its motion, and the kinetic energy should rise linearly with temperature (see Panel 1.3). But from our earlier results (eqns. (1.4) and (1.6)) and, assuming no defect scattering,

$$\mu_e \propto \tau_{\mathrm{L}} = \lambda \Big/ \sqrt{(\overline{c^2})} \propto T^{-3/2}.$$

The last step follows because the thermal velocity $\sqrt{(\overline{c^2})}$ is proportional to $T^{1/2}$.

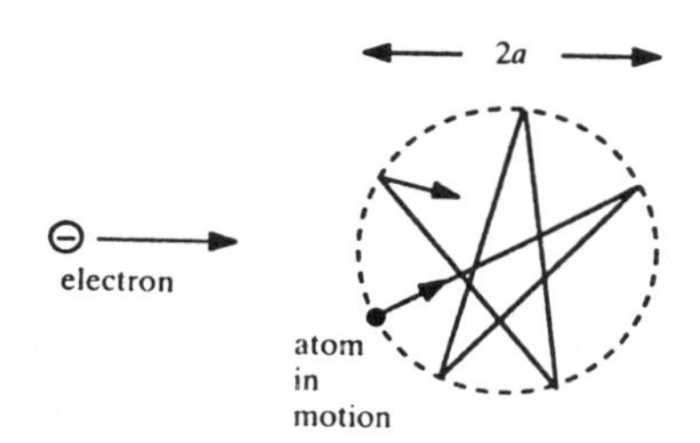

Fig. 1.10 An atom oscillating about a fixed position with an amplitude a presents an oncoming electron with a 'target' of area a^2.

This simple theory is not well obeyed in practice, as can be seen from Fig. 1.9. At low doping levels, where collisions with impurity atoms can be neglected, experiments show that μ_e is porportional to $T^{-3/2}$ in germanium, as expected. But under similar conditions in silicon, we find experimentally, to a close approximation, that $\mu_e \propto T^{-5/2}$.

In contrast, the mean time τ_{I} between collisions with impurities generally rises with temperature, in a rather complicated way.

1.7 Measurement of mobility and determination of carrier sign: the Hall effect

In 1897 J. J. Thompson determined the sign of the electron's charge by observing the deflection of an electron beam in both magnetic and electric fields. In a similar way it is possible to distinguish between holes and electrons. At the same time the carrier concentration can be determined from a measurement of the Hall effect, as explained below. This enables us to find the electron or hole mobilities, from a separate measurement of the conductivity of either *n*-type or *p*-type material.

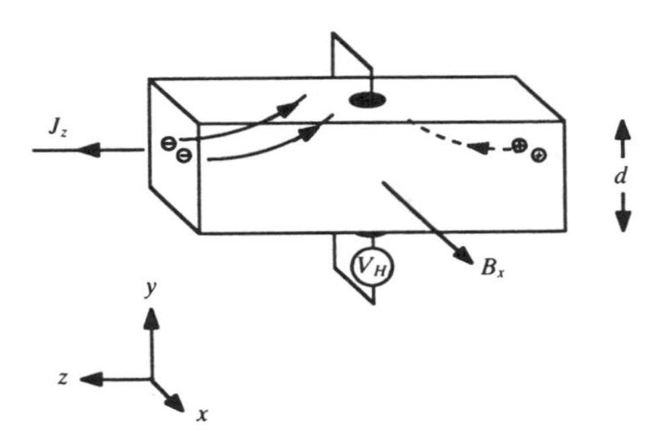

Fig. 1.11 Illustrating the Hall effect in a semiconductor.

Consider the semiconductor bar in Fig. 1.11 to be carrying a current density J_z in the z direction, due to electrons moving in the opposite direction. A magnetic flux B_x is appplied along the x axis. Each electron moving at velocity $-v_z$ should experience a force along the positive y axis, of magnitude

$$F_y = B_x(-e)(-v_z) = B_x e v_z \tag{1.8}$$

It will therefore accelerate in the y direction, towards the upper face of the bar in the figure. The upper face thus becomes charged negatively with an excess of electrons, while a balancing positive charge appears on the lower face. The resulting potential difference between these two faces — the HALL VOLTAGE — can be measured with a high-impedance voltmeter.

We shall now show that the Hall voltage is inversely proportional to the carrier concentration.

Let us begin by assuming that all electrons move with the same drift velocity $v_z = -\mu_e E_z$, where μ_e is the mobility of an electron. This assumption is not correct, but to assume otherwise makes the calculation of the Hall voltage much more complicated. We also assume that the high-impedance voltmeter draws no current, so that when the voltmeter reading is steady, the force on the electrons due to the magnetic field is just balanced by an electric field E_y which is due to the charges established on the top and bottom faces of the bar. This balance of forces is expressed in the equation

$$B_x e v_z + (-e)E_y = 0 \tag{1.9}$$

But since $J_z = nev_z$ we find by substitution for v_z in eqn. (1.9):

$$E_y = B_x v_z = \frac{B_x J_z}{ne}$$

The voltage V_H measured on the voltmeter, across the bar of thickness d, is related to the field E_y by $E_y = -V_H/d$, so that we find

$$\frac{V_H}{B_x J_z d} = \frac{-E_y}{B_x J_z} = -\frac{1}{ne}$$

The quantity on the left hand side of this equation is called the HALL COEFFICIENT, symbol R_H.

Thus

$$R_H = -\frac{1}{ne} \quad \text{(electrons)} \tag{1.10}$$

If this experiment is carried out on a *p*-type semiconductor, the Hall voltage is found to have the *opposite sign*, showing that the upper face of the bar in Fig. 1.11 acquires an excess *positive* charge.

We can interpret this in terms of the motion of carriers with positive charges — the 'holes' introduced earlier. Since positive charge carriers must move in the *same* direction as the current, their velocity v_z has a *positive* sign. The force F_y, being proportional to the product of charge and velocity, remains positive in sign, i.e. it is upward in Fig. 1.11. Thus the voltmeter reading is reversed in sign, as observed. Moreover, a measurement of the Hall voltage enables the concentration of carriers to be determined and compared with the doping density, while the effective mobility μ_h of holes can be found if the conductivity is also measured.

We can show this in exactly the same way as we did for electrons. The force balance equation for charges of positive sign and concentration p is

$$B(+e)\left(\frac{J_z}{pe}\right) + (+e)E_y = 0$$

and the Hall coefficient is

$$R_{\mathrm{H}} = +\frac{1}{pe} \quad \text{(holes)} \tag{1.11}$$

The results of measurements of R_{H} on p-type semiconductors lead to the conclusion that around room temperature, $p = N_A$, as stated earlier.

By measuring the Hall coefficient the carrier density n or p can thus be found. Since conductivity $\sigma = ne\mu_e$ or $\sigma = pe\mu_h$, depending on doping, we can find the mobility:

$$\mu_e = \sigma|R_{\mathrm{H}}| \tag{1.12}$$

Mobility determined this way is called the HALL MOBILITY, and is slightly in error owing to the neglect of the spread in carrier velocities.* Hence mobilities found from other experiments may not give precisely the same value.

Mobilities measured at 300 K for various semiconductors, and corrected for velocity spread, are given in Table 1.2.

Note that the electron mobility is always greater than the hole mobility — by about $2\frac{1}{2}$ times in silicon.

---

*Remember that we assumed at the outset that all carriers had the same drift velocity v_z. The correct equation properly adjusted for the spread in velocities is

$$R_{\mathrm{H}} = \frac{r}{pe} \quad \text{or} \quad \frac{-r}{ne} \tag{1.13}$$

where $r \simeq 1.0$ to 1.5, depending on the way the crystal lattice affects the velocity spread. Since determination of r is tricky, we often assume it to equal unity.

Table 1.2

Semiconductor	μ_e m^2/Vs	μ_h m^2/Vs
Si	0.135	0.048
Ge	0.39	0.19
GaAs	0.85	0.048
InSb	8.0	0.02
InAs	2.3	0.01

1.8 Interpretation of the hole mobility

Measurements of the concentration and mobility of holes raise the question of how we are to imagine the charge motion in a *p*-type semiconductor. The measurements of mobility show a temperature dependence and an impurity dependence similar to those of electrons: we are forced to conclude that holes appear to have thermal velocities like free electrons, and 'collide' with both thermal vibrations (phonons) and with impurity atoms.

We have likened the hole to a bubble in a lattice of pipes. If the 'bubble' is continually in motion with a thermal velocity comparable to that of the free electrons, it means that *the electrons in the surrounding bonds must be in rapid thermal motion, too.* There is, indeed, no reason why this rapid interchange of electrons within bonds should not occur even when no holes are present. This motion is not directly observable, for no current flows as a result, except in the presence of holes (a filled set of bonds, as we have seen, carries no current).

The electrons in the bonds must therefore be imagined as having random velocities arising from thermal energy in the solid. Holes move around with similar velocities to those of the bonding electrons surrounding them: they change direction when the neighbouring electrons change their direction. We can treat each hole as if it were a true particle carrying positive charge and with average thermal energy $\frac{3}{2}kT$. It acquires drift velocity v from an electrical field, and the rate of acceleration $\dot{v}$ is proportional to the force $+eE$ exerted by the field E, as expressed by the equation

$$\dot{v} = eE/m_h$$

Here m_h is the 'effective mass' of the hole — it can simply be regarded as a constant of proportionality between force and acceleration. The effective mass, coupled with the mean time τ between 'collisions' (i.e. change of direction) determine the hole mobility in exactly the same way as they do for free electrons:

$$\mu_h = \frac{e\tau}{m_h} \quad \text{and} \quad \tau = \frac{1}{\tau_I} + \frac{1}{\tau_L}$$

For practical purposes, it is sensible from now on to forget that a hole is an absence of an electron, and to treat it as a particle in its own right. We can therefore expect the hole mobility to vary with temperature and with defect density in a similar way to the electron mobility, as is indeed the case.

The drift velocity of both electrons and holes does not increase indefinitely as the electric field strength is raised. We can expect that, when the velocity eventually becomes comparable to the thermal velocity $\sqrt{(\overline{c^2})}$, the above analysis is no longer valid. The drift velocity then tends to a limit, called the SATURATION VELOCITY v_{sat}, which is very roughly equal to the thermal velocity in practice, and is close to 10^5 m/s in silicon at room temperature.

1.9 Total drift current

As a piece of semiconductor always contains both electrons and holes, the drift current density must be the sum of the drift currents due to both:

$$J = J_e + J_h = (ne\mu_e + pe\mu_h)E \tag{1.14}$$

In doped semiconductors, the drift current due to minority carriers can always be ignored (since $p \ll n$ or vice versa — see next section), so the above equation is primarily of use in intrinsic or nearly compensated semiconductors. (A doped semiconductor is COMPENSATED when $n = p$; this can be achieved by making $N_A = N_D$ if the doping levels are not too high.) Thus the carrier density n_i or p_i in intrinsic silicon may be calculated from the measured conductivity of compensated material, which is close to 4×10^{-4} S/m at 300 K:

$$\sigma_i = n_i e(\mu_e + \mu_h) \quad \text{since} \quad n_i = p_i$$

Hence in silicon at 300 K:

$$n_i = \frac{\sigma_i}{e(\mu_e + \mu_h)}$$

$$= \frac{4\times10^{-4}}{1.6\times10^{-19}(0.135+0.048)} = 1.4\times10^{16}\ \text{m}^{-3}$$

1.10 Minority carrier concentration: the *np* product

Thermal vibrations of the semiconductor lattice are continually generating small numbers of minority carriers by shaking free some of the electrons in bonds. This process creates each time a hole and an electron, at the expense of an amount of energy E_g, called the 'ENERGY GAP', which is related to the strength of a silicon–silicon atomic bond.

Generation of carrier pairs is balanced in thermal equilibrium by their disappearance by RECOMBINATION — when a free electron encounters a hole and 'falls into' it, giving up the energy E_g. The *rate* of generation G of pairs per unit volume and time thus equals the rate of recombination R per unit volume and time, i.e. $R = G$. Because the impurity level is so small, G is unaffected by it, and is thus independent of both majority and minority carrier concentrations. Since $R = G$, the recombination rate R must also be independent of doping — a fact we can use to advantage.

On the other hand, if we now model a process by which recombination could occur, we shall find that R depends on the product of n and p, and we are obliged to conclude that THE PRODUCT *np* IS INDEPENDENT OF DOPING. The argument proceeds as follows.

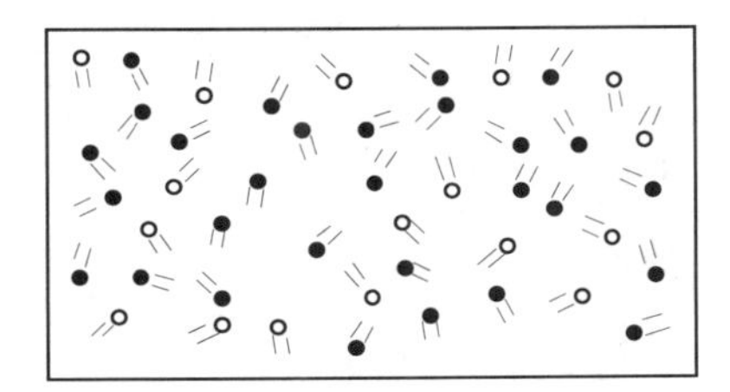

Fig. 1.12 The rate of collisions between black and white balls in random motion is proportional to the product of their concentrations.

The rate of recombination R must depend on the frequency with which a minority carrier (e.g. a hole) encounters a majority carrier (e.g. an electron). As both are in random (thermal) motion, we may think of them as like black and white pool balls in rapid motion on a pool table (Fig. 1.12). The rate of collision of *one* hole with electrons must be proportional to the concentration n of electrons. The rate of collisions of p holes with electrons is therefore proportional to the product np. Having encountered one another, the probability of recombination of hole and electron is independent of n and p. Hence the recombination rate R, is proportional to np, the constant of proportionality depending on carrier velocities (i.e. on temperature T) and on other concentration-independent factors. So we can write

$$R = np \times f(T)$$

where $f(T)$ is independent of n and p.

Now as $R = G$, and the generation rate G is not expected to depend upon doping at all, the product np must be independent of the values of n and p. In particular, this product in an intrinsic semiconductor, in which $n_i = p_i$ (cf. eqn. (1.1)) is equal to n_i^2, i.e.

$$np = n_i p_i = n_i^2 \tag{1.15}$$

Given n_i, this result can be used to find the minority carrier concentration in a sample with known majority carrier density. This will be most useful when we come to study the behaviour of devices which depend on the presence of minority carriers.

For example, in an n-type semiconductor, in which $n = N_D$, we have a hole concentration p_n

$$p_n = \frac{n_i^2}{n_n} = \frac{n_i^2}{N_D} \quad (n\text{-type}) \tag{1.16}$$

and in a p-type sample:

$$n_p = \frac{n_i^2}{p_p} = \frac{n_i^2}{N_A} \quad (p\text{-type}) \tag{1.17}$$

(Note the subscripts, which indicate the *type* of semiconductor referred to.)

It is notable that since n_i, which is about 10^{16} in silicon, is much less than N_A or N_D we have

$$p_n, n_p \ll n_i$$

In other words, the minority carrier density is always much less than the intrinsic concentration.

For example in a lightly doped silicon sample with $N_D = 10^{21}$ m^{-3}, we find $p_n = n_i^2/N_D = 10^{11}$ m^{-3}.

So a volume of 10^3 µm^3 — about the size of a transistor in an integrated circuit — would contain on average 10^{-4} minority carriers! It is clear that they can normally be neglected.

Germanium is different. For reasons we shall meet in the next chapter, its intrinsic carrier density is much higher, at about 2.5×10^{19} m^{-3}. So the minority carrier density for a given doping level is much greater than in silicon. This has relevance to its suitability for diodes and transistors, as we shall see.

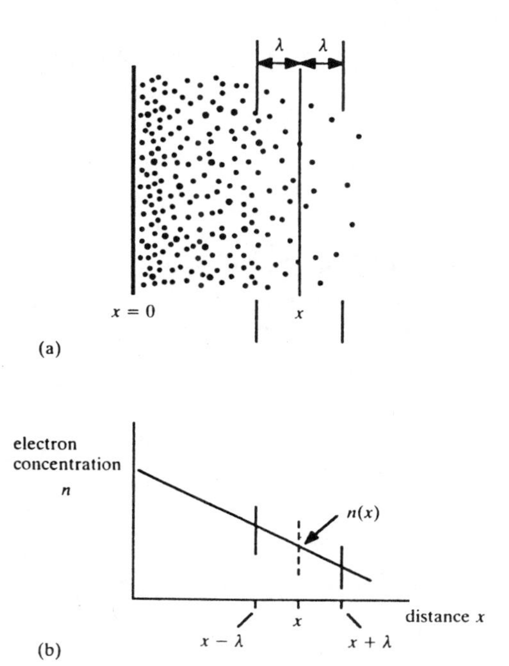

Fig. 1.13 A gradient in the concentration of electrons leads to diffusion.

1.11 Diffusion of electrons and holes

Because free electrons behave like a gas, it should not be surprising to find that electrons will DIFFUSE, in the way that gas molecules do, from a region of high concentration to one of low concentration. When we come to study devices in Chapters 3–5, we shall often find situations in which the minority carrier concentration is a function of position. The simplest case is depicted in Fig. 1.13, where it depends only on one coordinate x.

Diffusion is a consequence of the *random thermal motion* of electrons, and does not result from forces on them due to electric or magnetic fields. Using Fig. 1.13, we now show that the resulting rate of flow of electrons is directly proportional to the concentration gradient.

Consider the flow of electrons across a plane of unit area at right angles to the x-axis, cutting it at the point x where the concentration of electrons is $n(x)$. Electrons to left or right of this plane have equal probabilities of crossing the plane. But because there are more of them on the left than to the right, there is a net flow towards the right. Such electrons underwent their last collisions before crossing at an average distance λ_x from the plane (λ_x is related to the mean free path). The concentration of electrons at a distance λ_x to the left is $(n(x) - \lambda_x dn/dx)$. So the flux, i.e. the number crossing unit area of the plane in unit time, is

$$\frac{1}{2}\bar{u}\left(n(x) - \lambda_x \frac{dn}{dx}\right)$$

where $\bar{u}$ is the root mean square velocity parallel to the x-axis, and the factor of $\frac{1}{2}$ is present because only one half of the electrons are moving toward the right at any instant. Since the net flow F across the plane is the difference of the fluxes from the left and right, it can be written

$$\begin{aligned} F &= \frac{\bar{u}}{2}\left(n(x) - \lambda_x \frac{dn}{dx}\right) - \frac{\bar{u}}{2}\left(n(x) + \lambda_x \frac{dn}{dx}\right) \\ &= -\bar{u}\,\lambda_x \frac{dn}{dx} \end{aligned}$$

The flow is proportional to the concentration gradient dn/dx, so we usually write

$$F = -D_e \frac{dn}{dx} \tag{1.18}$$

where D_e is called the DIFFUSION COEFFICIENT of electrons, given by

$$D_e = \overline{u}\,\lambda_x \tag{1.19}$$

Note that the negative sign in eqn. (1.18), coupled with the negative gradient in Fig. 1.13(b), implies that the flow is to the right in the figure. The associated electric current is in the opposite direction, since each electron carries a charge $-e$. The current density, J, resulting from eqn. (1.18) is thus

$$J = -eF = +eD_e \frac{dn}{dx} \tag{1.20}$$

Holes can of course diffuse in exactly the same way, obeying an equation exactly similar to eqn. (1.18). But since each carries a positive charge e, the current which results flows in the same direction as the holes, i.e.

$$J = -eD_h \frac{dp}{dx} \tag{1.21}$$

Now consider what happens when both a concentration gradient *and* an electric field are present. Provided that the randomness of thermal motion is preserved, the net flow is the sum of the two separate effects. Adding the drift and diffusion currents then gives an equation for the total current due to electrons:

$$J_e = ne\mu_e E + eD_e \frac{dn}{dx} \tag{1.22}$$

The total hole current is similarly

$$J_h = pe\mu_h E - eD_h \frac{dp}{dx} \tag{1.23}$$

These equations will be particularly useful when discussing *p–n* junctions in Chapter 3.

1.12 Einstein's relation between D and μ

The fact that both the mobility μ and the diffusion coefficient D are related to the mean free path suggests that there is a connection between them. We shall now show that they are directly proportional to one another.

The mean distance λ_x between the last collision of an electron and a neighbouring plane is actually equal to the mean distance between collisions measured parallel to the x-axis. Although it might seem that λ_x would be smaller, this is not correct. An arbitrarily chosen plane is more likely to intersect longer paths than shorter ones, so that λ_x must be greater than half the x-component of the mean free path. Indeed, if one measures the mean distance to the next collision from any *randomly* chosen point on an electron's path, it equals the mean free path! This is because the probability of collision is independent of previous history.

Now, because $\lambda_x = \lambda_y = \lambda_z$ and $\lambda_x^2 + \lambda_y^2 + \lambda_z^2 = \lambda^2$, we have $\lambda_x^2 = \lambda^2/3$, where λ is the mean free path defined earlier. In a similar way, the mean square x-component of velocity is $\overline{u^2} = \overline{c^2}/3$.

Using these results in eqn. (1.19) shows that

$$D = \frac{\lambda}{3}\sqrt{\left(\overline{c^2}\right)}$$

Eqns (1.4) and (1.6) when combined give the following expression for the mobility:

$$\mu_e = \frac{e\tau}{m_e} = e\lambda \Big/ \sqrt{\left(\overline{c^2}\right)}$$

Dividing the two equations for D_e and μ_e and using eqn. (1.2) results in the equation

$$\frac{D_e}{\mu_e} = \frac{kT}{e}$$

which is called EINSTEIN'S RELATION. It applies equally well to holes as to electrons, and enables us to find D from the mobility values in Table 1.2. In spite of the approximations made here in deriving expressions for   and D, their ratio given by this relation is exact.

Summary of new terminology

The following are additional to those already collected on pages 4 and 12.

The HALL COEFFICIENT R_H is defined as the Hall field strength per unit of current and of magnetic flux density.

The EFFECTIVE MASS m_h of a hole is the constant of proportionality relating force and acceleration.

RECOMBINATION is the process by which a free electron disappears by filling a free hole.

The ENERGY GAP is the amount of energy needed to create a free electron and a free hole by 'breaking' an atomic bond, a process called GENERATION of hole-electron pairs.

A COMPENSATED semiconductor has $N_A = N_D$, giving a conductivity close to that of an INTRINSIC sample.

The DIFFUSION COEFFICIENT D_e of electrons and D_h for holes is the constant of proportionality between flux F of particles and concentration gradient.

PROBLEMS

1.1 Give the meaning of the terms p-type and n-type semiconductor. Write down as many as you can of the ways in which they differ from an *intrinsic* semiconductor.

1.2 Using a copy of the periodic table, list all the elements which have the correct valency to act as (a) donors (b) acceptors.

1.3 List all elements with the right valency to be semiconductors. Find out all you can about their electrical properties.

1.4 A pure silicon crystal is doped with 5×10^{23} boron atoms per cubic metre. What is the concentration of holes at (a) room temperature (b) near the absolute zero of temperature? What is the proportion of Si atoms which are replaced by boron? The density of Si is 2.33 g/cm^3 and its atomic weight 28.09.

1.5 Find the resistivities and minority carrier concentrations in
(a) germanium doped with 5×10^{20} atoms/m^3 of antimony (Sb)
(b) silicon doped with 10^{22} atoms/m^3 of boron (B)
(c) germanium doped with 5×10^{21} atoms/m^3 of indium (In)
Assume $n_i = 2.4 \times 10^{19}$ m^{-3}, $\mu_e = 0.39$ m^2/Vs and $\mu_h = 0.19$ m^2/Vs in germanium.

1.6 A crystal of Si doped with boron has a measured conductivity of 115 Sm^{-1} at $T = 300$ K. Use the appropriate mobility to find the concentration of charge carriers.

1.7 A resistor in an integrated circuit has dimensions 10 μm × 2 μm × 100 μm long. The doping level is uniformly 10^{23} phosphorus atoms per m^3, and the mobility at this doping level is half the 'textbook' value. Find the resistance at a temperature of 300 K.

1.8 Using $n_i = p_i = 1.4 \times 10^{16}$ m^{-3} for silicon at 300 K, find the concentration of holes in the sample of Problem 1.7, and show that the current they carry is negligible.

1.9 Give the meaning of the term *mobility* and explain its dependence on the frequency of collisions.

1.10 As indicated in Problem 1.7, a doping level of 10^{23} m^{-3} halves the mobility in *n*-type Si. Calculate the mean number of collisions per second made by an electron at this doping level, and compare it with the number in weakly doped Si. Explain the difference.

1.11 Distinguish between *thermal velocity* and *drift velocity* of carriers. Which is normally the larger? Which controls the mean time between collisions with the crystal lattice? How does a rise in temperature affect each velocity in a semiconductor?

1.12 Fig. P.1.1 shows an experimental set-up for measuring Hall voltages. The current I is held constant.

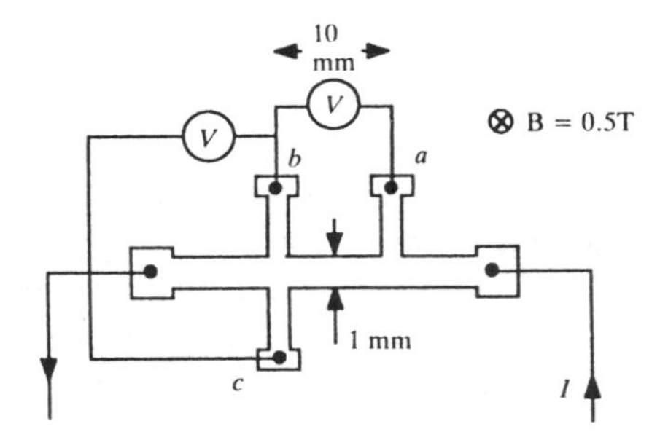

Fig. P.1.1

The voltages $V_a - V_b = 100$ mV and $V_b - V_c = 0$ μV are measured before applying a magnetic field downwards, normal to the diagram. Subsequently, with flux density $B = 0.05$ tesla, it is found that $V_b - V_c = 90$ μV. Find the semiconductor type, and the mobility.

1.13 If the current I in Problem 1.12 is 5 mA and the thickness of the sample is 0.1 mm, find the doping density in the semiconductor. Assume all donors or acceptors ionized, and uniform doping.

1.14 Explain why the expression for the Hall coefficient

$$R_H = -\frac{1}{ne} \quad \text{or} \quad \frac{1}{pe}$$

is only approximate.
In a Hall experiment two identical samples carrying the same current experience the same magnetic flux density of $0.2T$. It is found that the Hall field strength in one is 1.3 times the longitudinal field strength in the other. Find the mobility of majority carriers in the semiconductor.

1.15 Describe an experiment which can be used to distinguish whether a semiconductor is *p*-type or *n*-type. What quantitative information can the experiment give?

Chapter 2

Energy Bands and Carrier Statistics

Introduction

Hitherto we have used the 'bond model' of a semiconductor, and we have discussed the energies of free electrons and holes but little. There is a very useful alternative model, in which energy levels are paramount, and with which relatively simple expressions for the electron and hole concentrations can be derived. This is the *energy band* model, which will now be described in outline, before justifying it more carefully.

We have seen that free electron-hole pairs can be created in a semiconductor by giving some energy to an electron in a bond. This, and other features of a semiconductor, can be illustrated on an ENERGY LEVEL DIAGRAM, as in Fig. 2.1. The vertical axis represents the amount of energy possessed by an electron in a particular STATE: for example the lower horizontal line shows the energy E_v of a stationary valence electron in a bond. The upper line represents the energy level E_c of a stationary free (conduction) electron, and is higher in energy by an amount E_g, called the ENERGY GAP. It is not possible for an electron in an intrinsic semiconductor

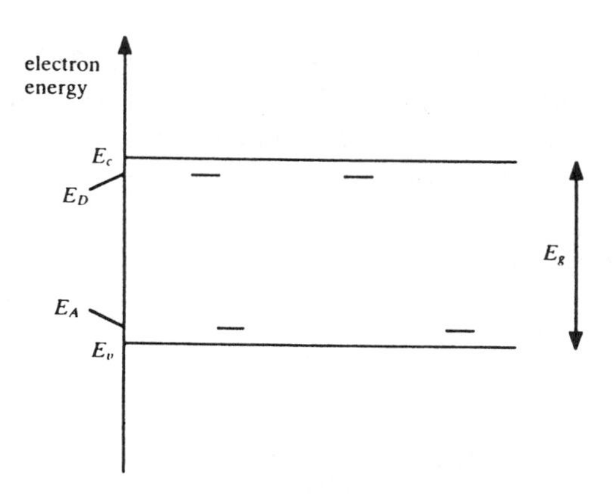

Fig. 2.1 Simplified electron energy level diagram for a semiconductor.

to possess an energy between the two levels, for E_g is the minimum energy needed to free an electron from a bond. This energy is the same everywhere, so the horizontal axis can be used to represent position in the crystal. Then the energy level associated with an electron bound to a donor atom is drawn as a *short* line marked E_D at a small distance below E_c, for this electron is localized in position, and requires but a small amount of energy to free it. Similarly, a short line just above E_v, and marked E_A, represents the energy level of an electron which has entered the half empty bond of an acceptor atom.

This model, together with a statistical treatment of the distribution of thermal energy among electrons, can be used to predict the concentration of electrons and holes at any temperature, given a knowledge of all the energies in the diagram. We shall see that, although this model is oversimplified, it gives remarkably accurate and useful predictions. In later chapters it will be used to discuss the behaviour of *p–n* junction diodes and transistors.

But first it is useful to review the origin of energy levels in isolated atoms, before explaining the modifications which occur when atoms are put together into a solid.

2.1 Energy levels in atoms

The sum of the kinetic and potential energies of an electron orbiting a nucleus can take only a set of discrete values; those for the simplest atom (hydrogen) are shown in Fig. 2.2. This fact is due to the wavelike properties exhibited by electrons, and cannot be explained by classical mechanics.

A simplified picture which helps in understanding this is that the electron wave, to be stable, must repeat itself exactly when followed around any circular path enclosing the nucleus. This is illustrated schematically in Fig. 2.3. It leads to the conclusion that an integral number of wavelengths must fit into the path, and is termed a 'quantization condition'.

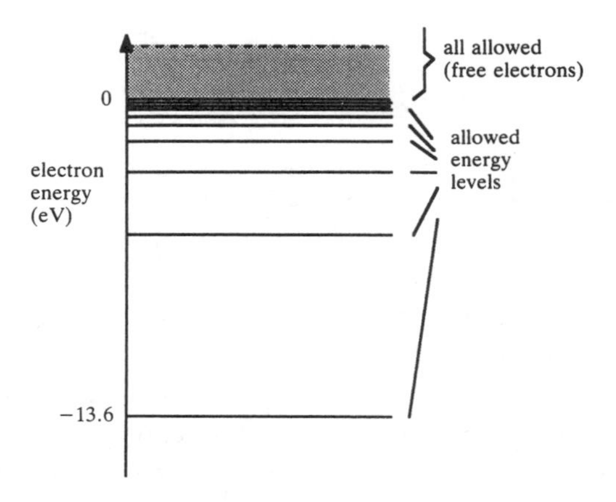

Fig. 2.2 Energy level diagram for a hydrogen atom.

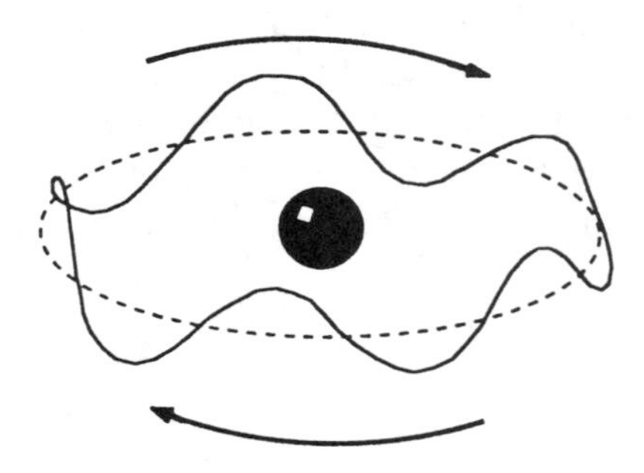

Fig. 2.3 Continuity of the electron wave around the nucleus results in an integral number of wavelengths fitting into any circular path.

Now the electron's wavelength λ is inversely related to its momentum p by De Broglie's relation: $\lambda = h/p$. The constant h is called Planck's constant, and has the value 6.6×10^{-34} Js. The quantization condition thus fixes the electron momentum to one of a set of discrete values, depending on the number of wavelengths fitting the circular path. The number, known as a 'quantum number', thus determines which of a set of discrete energies the electron has, since the energy is directly related to the electron's momentum. There are, in fact, three more quantum numbers to describe the electron's motion, but we need not consider them here.

Note that an electron freed from its parent nucleus can have any wavelength, and hence any kinetic energy. This explains the continuum of energy levels above the zero of energy in Fig. 2.2. When an atom contains many electrons, no more than two are found to exist simultaneously in the same 'orbit'. One of these electrons spins clockwise, the other anticlockwise, about its own centre of mass. So the above rule can be expressed by stating that only one electron is found in each 'state of motion' (orbit plus spin). This is known as PAULI's EXCLUSION PRINCIPLE, and a state of motion is usually called simply a STATE.

In the simplest case, when each state has a different energy we thus find electrons filling energy levels from the lowest up, one to each state. In the more general case, it is possible for two or more orbits, or states, to have the same energy. Thus a limited number of electrons are in the highest filled atomic energy levels, and these are the valence electrons, which take part in bonding an atom within a molecule or solid.

2.2 Energy levels in solids

By analogy with isolated atoms, we expect the energies of electrons in solids to have only certain allowed values. If we were to push many atoms closer and closer together to form a regular crystalline solid, the electron orbits in each atom would be modified by the attractive forces exerted on them by neighbouring nuclei. This causes the energy of each orbit to be shifted in such a way that the allowed energies form 'bands' of closely spaced levels, separated in energy from other allowed energy bands (see Fig. 2.4) by relatively large FORBIDDEN ENERGY GAPS. Within each band the allowed energies are so closely spaced that their separation is negligible ($<10^{-20}$ eV). This is because each level is derived from an energy

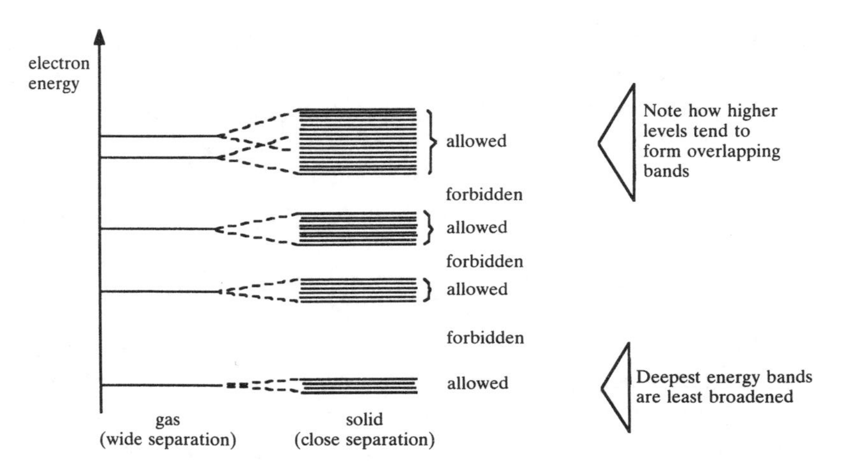

Fig. 2.4 Schematic energy level diagram of a solid, showing allowed bands and forbidden gaps.

level of one of the many atoms in the material. Thus in a cubic centimetre, which contains about 10^{23} atoms, the spacing of levels in a single energy band with total width 1 eV, is about 10^{-23} eV.

The higher atomic energy levels are associated with the outermost orbits, and they are spread into wider bands than are the lower levels, because the latter electrons are in tighter orbits around their parent nucleus and feel the influence of neighbouring nuclei much less. Hence the upper bands of levels tend to overlap with neighbouring bands, causing the energy gaps to disappear there.

Pauli's exclusion principle ensures that levels are still filled from the bottom up, and since there is a finite number of energy levels in each band, the highest filled level lies somewhere in one of the higher bands. Electrons in the higher bands move in 'orbits' which are so far out from the parent atom that they cannot be identified exclusively with it. These electrons are in fact in motion throughout the solid, and, in these upper energy bands, the horizontal axis in the energy level diagram can be used to represent position. Thus the valence electrons are shared, not just between neighbouring atoms, but throughout the whole crystal are said to be DELOCALIZED.*

*A more comprehensive justification for the energy band model than this necessarily brief treatment can be found in e.g. *The Physics of Solids* by C. A. Wert and R. M. Thomson, McGraw Hill, New York, 1970.

How, then, can we understand the differences between metal, semiconductor and insulator? The answer lies in the position of the highest filled energy level, in relation to the energy gaps between the bands, as we now discuss.

2.3 Metals, insulators and semiconductors

The main possibilities for the position of the highest filled level are indicated in Fig. 2.5. Each is associated with different electrical behaviour.

If the highest filled level lies *below* the top of a band of energies (Fig. 2.5(a) or (b)), an applied electric field will readily accelerate electrons, for they must move into empty higher energy levels as they gain energy from the field. An electron with a drift velocity of about 10 m/s has acquired about 3×10^{-10} eV of kinetic energy, and so moves up the energy axis by

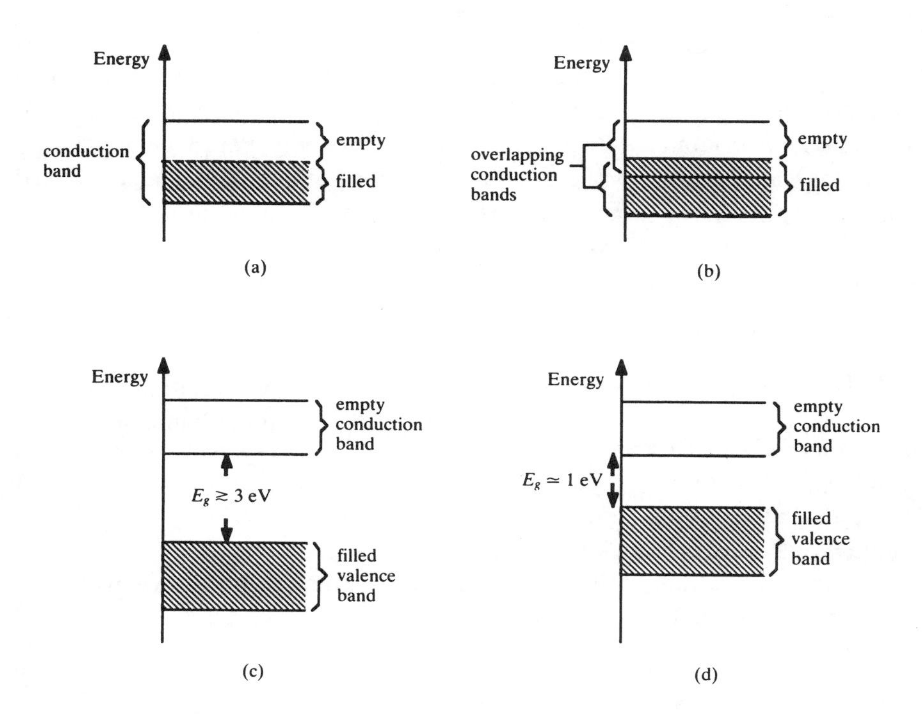

Fig. 2.5 Filled and empty energy levels in (a) and (b) metals, (c) an insulator and (d) a semiconductor.

just this amount. Figures 2.5(a) and (b) therefore represent metallic conductors: the former a simple, alkali metal, while the latter, a situation in which a filled band overlaps a partly filled band, is found in metals of higher valency, for example, aluminium.

If the highest filled level happens to lie exactly at the top of a band which does not overlap with another (as in Fig. 2.5(c) or (d)), any attempt to accelerate an electron with a small electric field must fail, since there are no suitable empty energy levels available for an accelerated electron to occupy. Thus a filled (and non-overlapping) energy band cannot conduct electricity; this throws a new light on the inability of valence electrons to conduct electricity without the presence of holes. The uppermost filled band is called the VALENCE BAND. The size of the energy gap between the valence band and the next band, called the CONDUCTION BAND determines whether the solid is an insulator or a semiconductor. If the gap is greater than 2 eV an insulator results, for the conduction band contains no electrons. This is illustrated in Fig. 2.5(c).

Figure 2.5(d) represents a semiconductor, in which the energy gap is small enough for a few electrons to have enough thermal energy at room temperature to cross the gap into the conduction band. So a semiconductor is merely an insulator with a relatively small energy gap. Conversely, an insulator can become a semiconductor, if the temperature can be raised sufficiently to cause some electrons to enter the conduction band.

Let us now consider how this energy gap might be measured, by discussing the absorption of light.

2.4 Light absorption and the energy gap

Just as thermal energy can cause electrons to cross from valence to conduction band, so other forms of energy can do the same. One of these is light energy, which Einstein showed comes in packets, or QUANTA, of minimum value hf, where h is Planck's constant again, and f the frequency of the light. The minimum packet, called a PHOTON, of green light with wavelength 0.5 μm, thus possesses an amount of energy:

$$E = hf = \frac{hc}{\lambda} = \frac{6.6\times10^{-34}\times3\times10^{8}}{0.5\times10^{-6}} = 4\times10^{19}\ \text{J}$$

This is almost 2.5 eV, and light of 0.5 μm wavelength cannot give up less energy to an electron in a solid. When such light is absorbed by a metal, we can depict on an energy level diagram (Fig. 2.6(a)) the electron transitions to higher energy levels, as each photon makes an electron move to an empty level 2.5 eV above its starting energy.

In the energy band diagram of an insulator in Fig. 2.6(b), on the other hand, there are no empty levels within 2.5 eV of any of the electrons in the valence band. Hence green light — and indeed all visible light — passes through the insulator without being absorbed. This explains the transparency of many insulating crystals such as diamond and quartz, when no impurities are present.

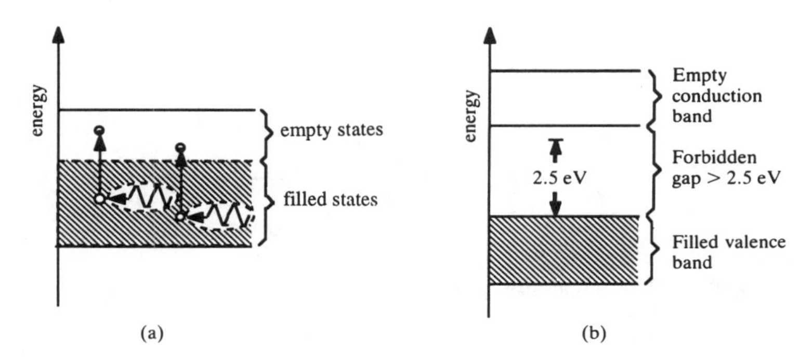

Fig. 2.6 A photon with 2.5 eV energy can give up its energy to an electron in a metal (a), but not in an insulator (b) with a large energy gap.

So the width of the energy gap of a material can be found by plotting the absorption of light against the wavelength, as is done for gallium arsenide in Fig. 2.7. The thickness of GaAs which absorbs a given fraction of the incident light is plotted logarithmically *down* the vertical axis of the figure, So increased absorption (represented by a rise *up* the vertical axis) occurs as the wavelength falls below about 0.85 μm, where the photon energy hf rises above the width of the energy gap, 1.4 eV. The energy gaps measured from such absorption edges, as they are called, are shown for several materials on the schematic spectrum in Fig. 2.8. Thus silicon and germanium, which absorb strongly in the visible part of the spectrum, look metallic. (Note that strong absorbers also reflect strongly: you can use this fact to explain why cadmium sulphide, with an energy gap of 2.5 eV, is coloured bright yellow.)

The generation of electron hole pairs by light forms the basis for a variety of different photodetector devices, one which is discussed in section 3.12.

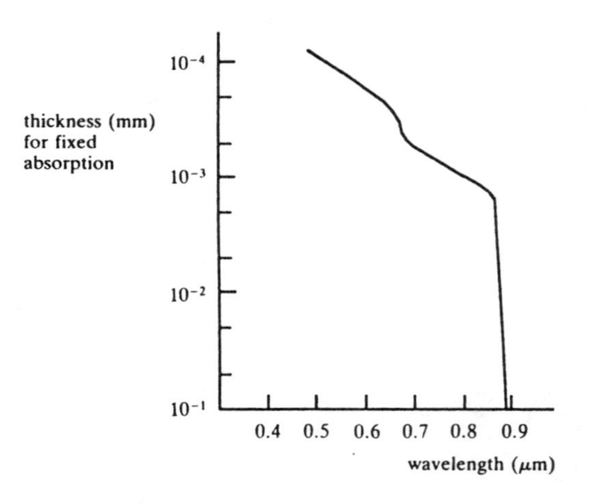

Fig. 2.7 The optical absorption in GaAs plotted against wavelength.

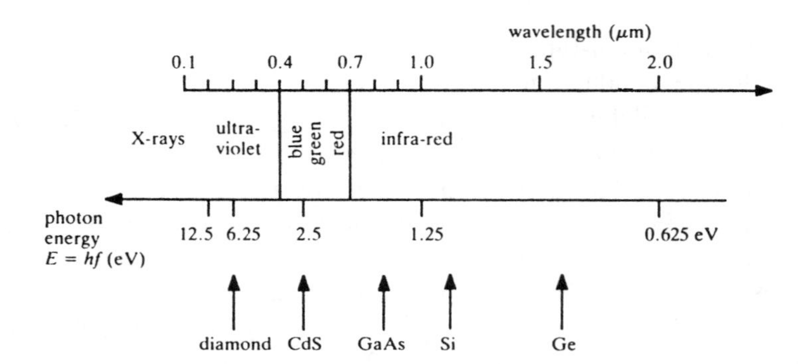

Fig. 2.8 The electromagnetic spectrum, showing wavelengths and photon energies at which various semiconductors begin to absorb radiation.

2.5 Impurity energy levels

The introduction of dopants to form an *n*-type or *p*-type semiconductor adds energy levels to the energy band diagram. The fifth electron, when attached to a donor atom, occupies an energy level E_D which is typically about 0.05 eV below E_c, the conduction band edge (see Panel 2.1 and Fig. 2.9).

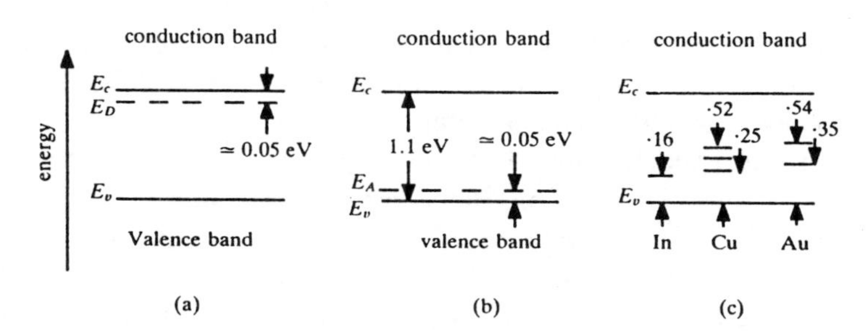

Fig. 2.9 Energy level diagrams for silicon containing (a) Group 15/V donor impurities (b) Group 13/III acceptor impurities (c) various other common impurities, energies in eV.

Likewise, an acceptor energy level in silicon lies about 0.05 eV above E_v, and represents the energy E_A of a captured electron bound to the acceptor. These energy levels are shown on the energy band diagrams in Fig. 2.9(a) and (b), while (c) shows the energy levels associated with the bound outer electrons of a few other impurities in silicon.

The energy level diagram alone does not explain why nearly all donor energy levels in an *n*-type semiconductor are vacant, having lost their electrons to the conduction band. To understand this we must find the proportion of available electron states which are occupied. This can be calculated by statistical means: we need an expression for the probability $f(E)$ that an electron state of energy E is occupied. This is discussed in the next section. We can then use the probability $f(E)$ and the number of available free electron states N_c per m^3 at energy E_c to express the concentration n of free electrons as the product:

$$n = N_c f(E_c)$$

Likewise, we can also calculate the proportion of *empty* states having the valence energy E_v — it is just $(1 - f(E_v))$. By multiplying this by the

PANEL 2.1

Donor energy levels

The fifth electron of a donor impurity is attracted to it by an electrostatic force. To estimate how much energy is needed to free the electron, it is possible to use Bohr's theory — first used for the hydrogen atom — which assumes that the captured electron orbits the positively charged nucleus in a circle of radius r. As illustrated below, the orbit is embedded in the semiconductor.

The inward acceleration of the orbiting electron is due to the force of attraction $e^2/4\pi\varepsilon r^2$, which is reduced below the force experienced in a hydrogen atom by the high permittivity ε of the semiconductor. As a result, the radius r is considerably enlarged, which further reduces the energy needed to ionize the donor atom. This energy can be shown to vary inversely as ε^2, so that in silicon, for which $\varepsilon = 11.7\varepsilon_0$, it is expected to be 11.7^2 times smaller than the ionization energy of a hydrogen atom. The latter is 13.6 eV, so we anticipate the donor ionization energy to be about 0.1 eV in silicon, and a similar amount in germanium. Measurements show that this simple model overestimates the energy by about a factor of 2 in silicon, i.e. it is about 0.05 eV. (The overestimate is even bigger in germanium.) Since the freed electron in the conduction band has the energy E_c, the level E_D of the bound electron lies 0.05 eV below E_c in silicon.

A similar model can be used to calculate the energy needed to free a hole 'orbiting' an acceptor atom, with an identical result to the above.

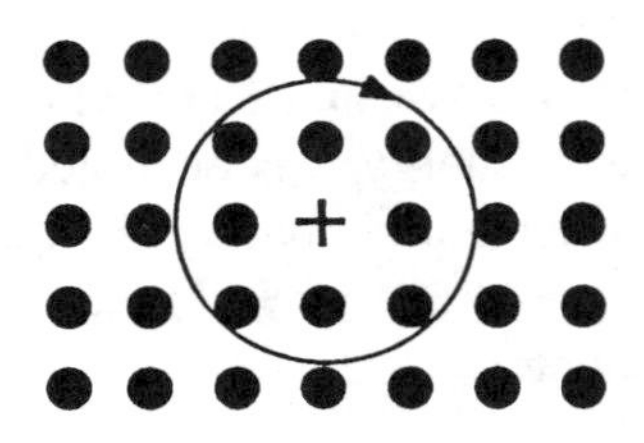

concentration of available valence electron states N_v the concentration of holes is similarly found, i.e.

$$p = N_v(1 - f(E_c))$$

After considering the form of the function $f(E)$ in the next section, the simple expressions above will be justified rather more carefully in section 2.11.

2.6 Probability of occupation of electron states: the Fermi function

Since electrons (or other particles) are in random motion, each would be found with equal probability in any energy level, were it not for two constraints:

(a) that the total energy of all particles is fixed at a given temperature by the amount of heat energy in the system, and
(b) no two electrons may be in the same state (Pauli's exclusion principle again).

To illustrate the principles involved, consider a simple example of a hypothetical system having states equally spaced in energy: zero energy in the lowest, and 1, 2, 3, ... units in successively higher levels. Let there be three electrons present, with a total of five units of energy among them, only one electron being allowed in each level. Two possible arrangements are shown in Fig. 2.10; they are the only two arrangements which have the same total energy. Interchange of a pair of electrons does not give a different arrangement, since electrons are indistinguishable. Hence these two arrangements are equally probable, and each must occur for 50 per cent of the time. The probability of finding the levels with 1, 2, 3 or 4 energy units occupied is thus 0.5. The same principles can be applied with a much larger number of electrons and energy levels to find the probability accurately as a function of energy. Panel 2.2 shows a larger example which you can check

energy of state	0	1	2	3	4	5	
	●		●	●			Total energy 0 + 2 + 3 = 5
	●	●			●		Total energy 0 + 1 + 4 = 5

Fig. 2.10 Possible ways of arranging three electrons with five units of energy among states with unit energy spacing.

PANEL 2.2

The fifteen possible arrangments of 11 electrons with 62 units of energy in equally spaced levels are shown in the chart. The number of arrangements in which each is occupied is shown in the bottom row. Expressed as fifteenths, these numbers represent the probabilities that the levels are occupied, and are plotted on Fig. 2.11 for comparison with the Fermi function, eqn. (2.1).

To see that the shape is independent of the spacing of levels, consider opening an energy gap by removing any two adjacent states, e.g. levels 10 and 11. This reduces the number of possible arrangements. The remaining arrangements include only those with *no* electrons in boxes 10 and 11, and no further arrangements are possible (otherwise they would have been included already; i.e. there is no *new* way to redistribute the 1 or 2 particles in boxes 10 and 11 in the rejected arrangements). The fractions in the final row are little changed as a result: if the numbers of electrons, and hence of arrangements, were made very much larger, the changes would become negligibly small.

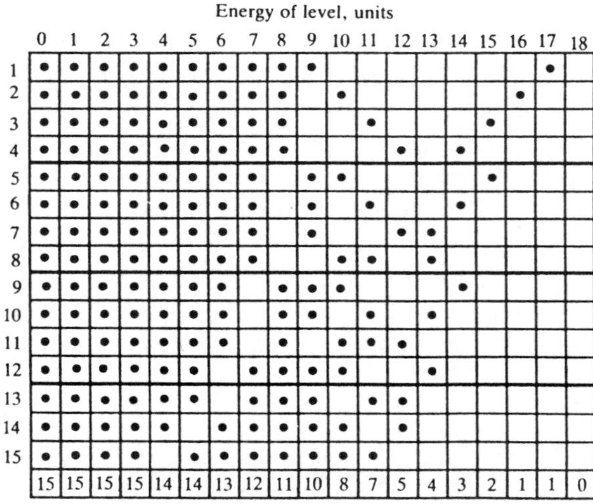

Energy of level, units

	0	1	2	3	4	5	6	7	8	9	10	11	12	13	14	15	16	17	18
1	•	•	•	•	•	•	•	•	•	•								•	
2	•	•	•	•	•	•	•	•	•		•						•		
3	•	•	•	•	•	•	•	•	•			•				•			
4	•	•	•	•	•	•	•	•	•				•		•				
5	•	•	•	•	•	•	•	•		•	•					•			
6	•	•	•	•	•	•	•	•		•		•			•				
7	•	•	•	•	•	•	•	•		•			•	•					
8	•	•	•	•	•	•	•	•			•	•		•					
9	•	•	•	•	•	•	•		•	•	•				•				
10	•	•	•	•	•	•	•		•	•		•		•					
11	•	•	•	•	•	•	•		•		•	•	•						
12	•	•	•	•	•	•		•	•	•	•			•					
13	•	•	•	•	•	•		•	•	•		•	•						
14	•	•	•	•	•		•	•	•	•	•		•						
15	•	•	•	•		•	•	•	•	•	•	•							
	15	15	15	15	14	14	13	12	11	10	8	7	5	4	3	2	1	1	0

No. of arrangements in which energy level is occupied

by hand, and compare with the equation given below which has been deduced analytically for a *very* large number N of electrons with total energy W.

The problem is that of finding the function $f(E_i)$ that gives the probability that the ith state, having energy E_i, is occupied, subject to the two constraints:

$$\sum_{i=1}^{\infty} f(E_i) = N$$

and

$$\sum_{i=1}^{\infty} E_i f(E_i) = W$$

The statistical argument presented in physics textbooks shows that the function $f(E)$ has the form

$$f(E) = \frac{1}{1 + \exp\left[(E - E_F)/kT\right]} \tag{2.1}$$

where k = Boltzmann's constant, and E_F, called the FERMI ENERGY or FERMI LEVEL, is the value of E at which the probability $f(E) = \frac{1}{2}$. It also is the energy of the highest filled level when the temperature T is 0 K. The shape of the FERMI FUNCTION, as $f(E)$ is called, is illustrated in Fig. 2.11

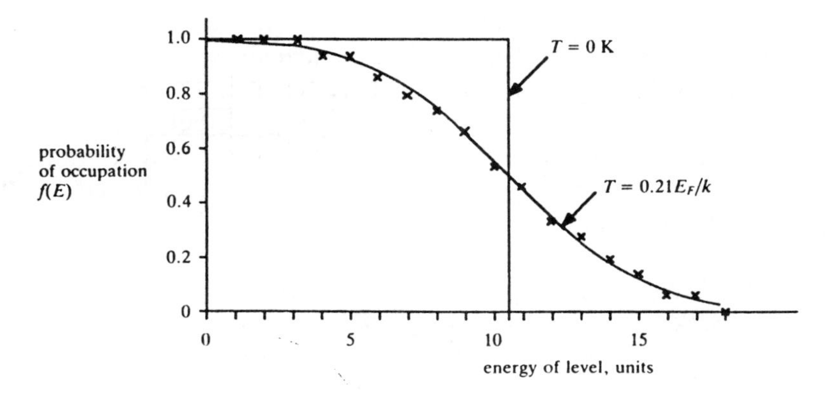

Fig. 2.11 The Fermi function, plotted against energy for two different temperatures. The crosses show the values predicted in Panel 2.2.

for the temperatures $T = 0$ and $T = 0.21\,E_F/k$. In order to grasp the nature of the result, it may help if you find the slope of $f(E)$ at the point $E = E_F$ by differentiating eqn. (2.1), and use the result to sketch in the approximate shape when $T = E_F/10k$ on Fig. 2.11.

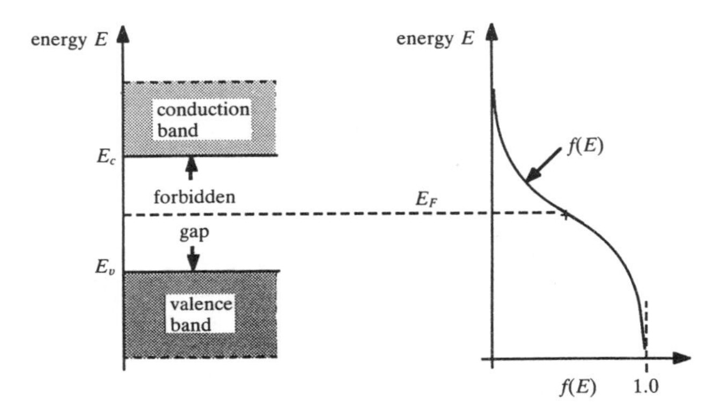

Fig. 2.12 Illustrating the use of the Fermi function to calculate electron concentration in an intrinsic semiconductor.

2.7 Calculation of the free electron concentration: intrinsic semiconductors

The Fermi function is used in the following way. It is re-drawn (on its side) alongside the energy level diagram of a semiconductor in Fig. 2.12, with the Fermi level midway between E_c and E_v. Suppose we wish to find the concentration n_c of electrons having energy E_c. If the concentration of available electron states with this energy is N_c, then the concentration of electrons sought is N_c multiplied by the probability that each state is filled, which is just $f(E_c)$, i.e.

$$n_c = N_c f(E_c) = \frac{N_c}{1 + \exp\left(\dfrac{E_c - E_F}{kT}\right)} \tag{2.2}$$

Note from Fig. 2.12 that $E_c - E_F \simeq 0.5$ eV, and is therefore much greater than kT, which is about 1/40 eV at 300 K. Hence the exponential in eqn. (2.2) is much the greater term in the denominator, and we can neglect unity compared with it, resulting in

$$n_c = N_c \exp -\left(\frac{E_c - E_F}{kT}\right)$$

Now electrons in the conduction band are actually spread over all the energies above E_c. But note that, at an energy of $2kT$ above E_c, the Fermi function reduces by a factor of e^2, or nearly 10. Since $kT = 0.025$ eV at 290 K, the vast majority of free electrons occupy energy levels within about 1/20 eV of the edge E_c of the conduction band at normal temperatures.

This fact allows us to make a simple approximation by treating *all* the electrons as if they have the same energy E_c, adjusting the value of N_c so that it accounts for those electrons at higher energies. Thus the *total* free electron concentration is given to a good approximation by

$$n = N_c \exp -\left(\frac{E_c - E_F}{kT}\right) \tag{2.3}$$

where N_c works out to be about 2×10^{25} m^{-3} in silicon at 300 K, and differs rather little in other semiconductors.

In section 2.11 we shall show that this approximation can be justified more rigorously, and that it is quite accurate, both in intrinsic and doped semiconductors, provided only that the concentration of dopants is not too high.

In a similar way it is possible to find an expression for the hole concentration. The concentration p_v of holes at energy E_v is the concentration N_v of available electron states multiplied by the probability $[1 - f(E_v)]$ that each is empty, i.e.

$$p_v = N_v[(1 - f(E_v)] = N_v\left[1 - \frac{1}{1 + \exp\left(\frac{E_v - E_F}{kT}\right)}\right]$$

$$= N_v\left[\frac{\exp\left(\frac{E_v - E_F}{kT}\right)}{1 + \exp\left(\frac{E_v - E_F}{kT}\right)}\right]$$

Since $E_v - E_F$ is negative (see Fig. 2.12) and much larger in magnitude than kT, the exponential is much less than unity, and can be neglected in the denominator. This gives the excellent approximation

$$p_v = N_v \exp -\left(\frac{E_F - E_v}{kT}\right)$$

As with electrons, it is easy to argue that nearly all holes have energies close to E_v, and we can use a similar equation to represent the *total* hole concentration, using a suitably adjusted value of N_v, i.e.

$$p = N_v \exp -\left(\frac{E_F - E_v}{kT}\right)$$

In silicon, N_v has a value quite close to that of N_c, and, for convenience, we shall assume them to be equal.

Having drawn the Fermi function in Fig. 2.12 with E_F lying midway between E_c and E_v, we can now see that it represents an intrinsic semiconductor, for in this case

$$N_c \exp -\left(\frac{E_c - E_F}{kT}\right) = N_v \exp -\left(\frac{E_F - E_v}{kT}\right)$$

So that in this situation eqns. (2.3) and (2.4) show that $n = p$, as required. We conclude that the **Fermi level in an intrinsic semiconductor lies at the middle of the energy gap**.

The situation in doped semiconductors will be treated next.

Summary of symbols and terminology

E_c The energy of the lowest level in the CONDUCTION BAND.

E_v The energy of the highest level in the VALENCE BAND.

Electron state A state of motion available to a single electron having a fixed energy.

N_c The effective concentration of available electron states in the conduction band, at energy E_c.

N_v The effective concentration of available electron states in the valence band, at energy E_v.

$f(E)$ The FERMI FUNCTION, gives the probability that an electron state having energy E is occupied.

E_F The FERMI LEVEL, is that energy for which the FERMI FUNCTION has the value of 1/2.

E_g is the ENERGY GAP — the range of energy between E_c and E_v in which there are no electron states available in a semiconductor or insulator.

E_D is the energy of an electron in a DONOR ENERGY LEVEL, i.e. orbiting a donor atom.

E_A is the energy of an additional electron bound to an acceptor atom, i.e. in an ACCEPTOR ENERGY LEVEL.

2.8 Carrier concentration in doped semiconductors

Consider first an n-type material, in which we know that $n \gg p$. This means that the probability of finding electrons in energy levels in the conduction band is higher than that of finding empty energy levels in the valence band. This situation is depicted on the energy level diagram in Fig. 2.13(a), where it is seen that we achieve this end by putting the Fermi level E_F closer to E_c than to E_v.

Provided only that

$$\exp -\left(\frac{E_c - E_F}{kT}\right) \ll 1$$

which is satisfied when the value of $(E_c - E_F)/kT$ is above about 3, we can still use both eqns. (2.3) and (2.4) for the concentrations of electrons and holes (see the final section of this chapter).

Note that the product of these two equations leads at once to the result that the product np is independent of E_F:

$$\begin{aligned} np &= N_c N_v \exp -\left(\frac{E_c - E_F}{kT}\right) \exp -\left(\frac{E_F - E_v}{kT}\right) \\ &= N_c N_v \exp - E_g/kT \end{aligned} \tag{2.5}$$

where $E_g = E_c - E_v$ is the ENERGY GAP.

Because np is independent of the position of E_F relative to E_c and E_v, this product is independent of the doping level, which controls the value of $(E_c - E_F)$ in Fig. 2.13(a). The dependence of $(E_c - E_F)$ on doping can be seen from eqn. (2.3), since we know that the electron concentration n is equal to the donor concentration N_D. Thus

$$n = N_D = N_c \exp\left[-(E_c - E_F)/kT\right] \qquad (2.6)$$

Hence $E_c - E_F$ falls as N_D rises.

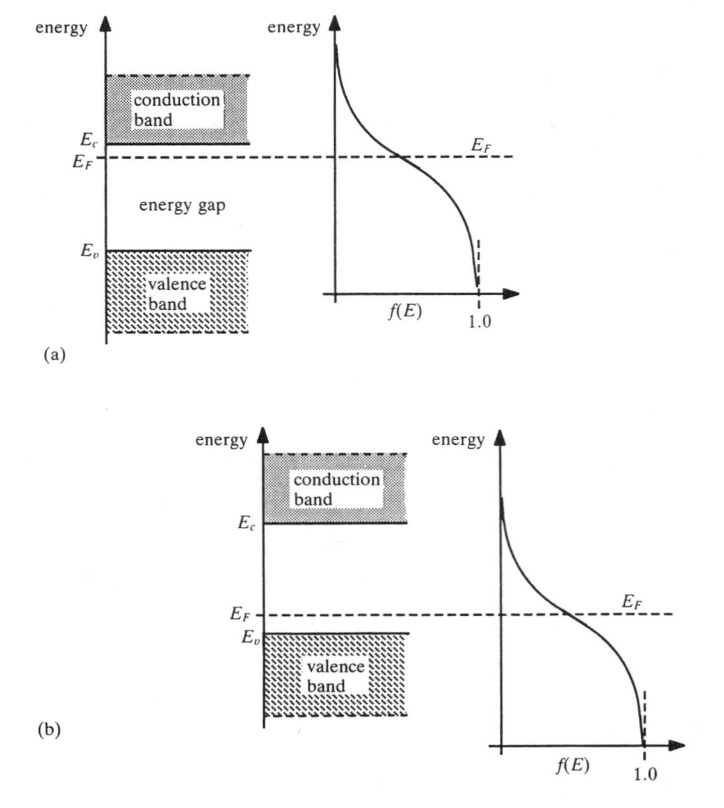

Fig. 2.13 Position of the Fermi level and Fermi function in (a) an n-type semiconductor (b) a p-type semiconductor.

Note that eqn. (2.6) does not imply that n changes strongly with temperature: rather, that $(E_c - E_F)$ adjusts itself to maintain n equal to N_D.*

Turning now to a p-type material, Fig. 2.13(b) shows how, when E_F is close to E_v, the value of $(1 - f(E_v))$ is greater than $f(E_c)$. In other words, the probability of finding empty levels at E_v is greater than that of finding electrons filling energy levels at E_c. Thus $p \gg n$ as required.

We conclude that, at normal temperatures, **E_F is near the top of the energy gap in an n-type semiconductor, but near the bottom of the gap in a p-type sample.**

Note that eqn. (2.5) explains how the intrinsic electron concentration n_i depends upon the energy gap. Knowing E_g and N_c we can find n_i: Thus in silicon, assuming $N_c = N_v = 2 \times 10^{25}$ m^{-3} and $E_g = 1.11$ eV, we have at $T = 300$ K:

$$n_i = \sqrt{N_c N_v} \exp - \frac{E_g}{2kT} = 1 \times 10^{16} \text{ m}^{-3}$$

In germanium, which has an energy gap of 0.67 eV while N_c and N_v are only slightly smaller, n_i is very much bigger, about 10^{19} m^{-3}. A small change in the energy gap thus produces an enormous change in n_i at room temperature.

2.9 Majority carrier density

We have not yet questioned why it is that the majority carrier concentration equals the doping density N_D or N_A. To answer this, let us consider an n-type semiconductor, and show first that the probability of finding a fifth electron attached to a donor atom is negligible, provided only that the doping is not too great.

Remember that, as shown in Fig. 2.13, the energy level E_D of an occupied donor atom lies just below E_c, by about 0.05 eV in silicon (see Panel 2.2). We can write down the fraction f of these donor levels which are filled, using the Fermi function in the normal way:

$$f(E_D) = \frac{1}{1 + \exp\left(\dfrac{E_D - E_F}{kT}\right)}$$

*This is only true for moderate changes in temperature. See section 2.9.

Thus as long as $\exp[(E_D - E_F)/kT] \gg 1, f(E_D)$ is small, and nearly all of the extra electrons will have left their parent donor atoms. The condition for this is simply that the Fermi level E_F should be more than about $3kT$ below the donor energy levels E_D. Equation (2.6) can be used to show that this is true at 300 K for all doping densities below about 10^{23} m^{-3}, when $N_c = 2 \times 10^{25}$ m^{-3} and $E_c - E_D = 0.05$ eV. Even at doping levels up to nearly 10^{25} m^{-3}, over half of the donor energy levels are vacant.

At all except the highest doping levels, then, all donor impurities can be assumed to be ionized. We may relate the free electron concentration to the donor concentration by balancing the charges carried by electrons, holes and ionized donor atoms. Since the semiconductor as a whole is neutral, the positive charges on donors and holes must be equal in quantity to the negative charges carried by free electrons, i.e.

$$N_D + p = n$$

Since $p \ll n$, we conclude from this equation that $n = N_D$.

In a p-type semiconductor, the energy level associated with an acceptor atom lies about 0.05 eV above the top of the valence band, and a similar argument to that used above shows that these energy levels are almost all occupied by electrons, provided again that N_A is not above about 10^{23} m^{-3} at 300 K. Again, consideration of the charge balance readily shows that, to a very close approximation, $p = N_A$.

2.10 Motion of free electrons and holes in energy bands

Since the horizontal axis in an energy band diagram such as in Fig. 2.13(a) and (b) represents spatial position, we can picture the motion of free electrons and holes across the diagram.

An electron promoted from the valence band into the lowest available energy level in the conduction band has no kinetic energy. It is 'just' free from the system of bonds, and must rise above the energy level marked E_c if it accelerates. We shall discuss its motion with the aid of Fig. 2.14. When a voltage is applied across the crystal, the energy band diagram takes on a tilt. This is because the potential energy of an electron is higher on that side of the crystal which is at the more negative potential, by an amount eV, where V is the potential difference. Now consider the motion of an electron sitting at a point A, in the energy level E_c. The electric field, represented now by the slope of the energy levels, causes it to move to the right, gaining kinetic energy while losing potential energy. Since its total energy is constant, the motion is represented by the horizontal line AB. If at B it collides with a defect or a phonon, it gives up the kinetic energy it gained, following the path BC on the diagram. Naturally the scale of events has been magnified in Fig. 2.14, but the overall effect is that the electron moves down the potential 'hill' created by the applied voltage, in a series of very small steps.

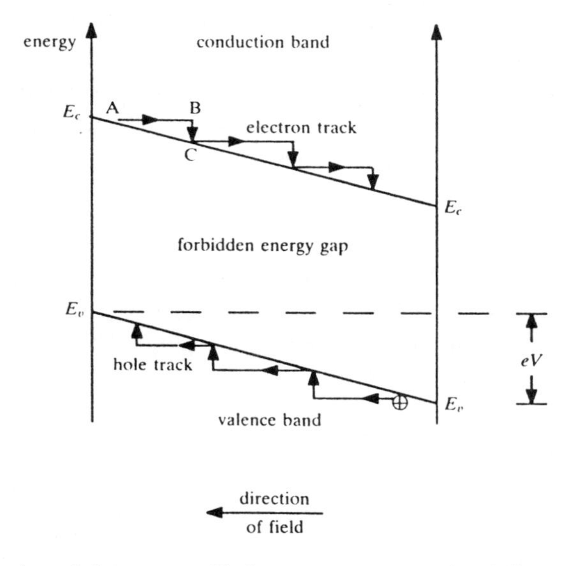

Fig. 2.14 The motion of electrons, and holes, across an energy level diagram in the presence of an electric field.

The motion of holes in the valence band can be discussed in a similar way. First we note that an electron at the *top* of the valence band, in the level E_v, is rather surprisingly, at rest. It has much potential energy, but no kinetic energy. Thus a hole at rest is simply a vacant level at E_v — we represent it by a circled plus sign in Fig. 2.14. So an accelerating hole moves *away* from E_v, *down* the scale of *electron* energy, as might be expected of something which behaves like a positively charged particle.

The motion of a hole on a diagram when an electric field is present is therefore an upside-down version of the free electron's zigzag. The hole moves *up* the potential hill, floating to the highest available point in the valence band, just as the bubble in the water pipe analogy in Chapter 1 floats upward in the gravitational field.

2.11 The energy distributions of electrons and holes

(*May be omitted at first reading*)
To calculate correctly the total concentration of electrons in the conduction band, it is necessary first to consider the concentration of free electrons with energies between values E and $E + dE$. This can be written as a product of the number of available states dN in the energy range dE and the Fermi function $f(E)$:

$$dn = f(E)dN \tag{2.7}$$

Now both dn and dN must depend on the energy E, so they will be written as $dn(E)$ and $dN(E)$ in future. The number of states $dN(E)$ must be proportional to the infinitesimal width dE of the range of energies, so we shall write

$$dN(E) = S(E)dE$$

from which it can be seen that

$$S(E) = \frac{dN(E)}{dE} \tag{2.8}$$

The function $S(E)$, called the DENSITY OF STATES, gives the concentration of states per unit of energy E. Its value is zero within the energy gap, where no electron states exist. In the conduction band, $S(E)$ rises, from the value zero at the energy E_c, in the way illustrated in Fig. 2.15(a). Over the range of energies just above E_c, the density of states has been shown to obey the equation

$$S(E) = A(E - E_c)^{1/2} \tag{2.9}$$

where the constant A depends only on the effective mass* m_e of electrons and on Planck's constant. The concentration $dn(E)$ of free electrons with energies between E and $E + dE$ is now found from eqns. (2.7) and (2.8):

$$dn(E) = f(E)S(E)dE \tag{2.10}$$

Now integrate $dn(E)$ over the whole conduction band, to obtain an expression for the total free electron concentration:

$$n = \int_{E_c}^{\infty} dn(E) = \int_{E_c}^{\infty} S(E)f(E)dE \tag{2.11}$$

By substituting the expressions for $f(E)$ and $S(E)$ given in eqns. (2.1) and (2.9) respectively, we find that

$$n = A\int_{E_c}^{\infty} \frac{(E - E_c)^{1/2}dE}{1 + \exp(E - E_F/kT)} \tag{2.12}$$

(Strictly, the expression (2.9) for $S(E)$ is invalid at higher energies, but the error incurred is negligible, since $f(E)$ becomes extremely small at those energies. Most of the electrons have energies within $3kT$ of E_c, as we shall show.)

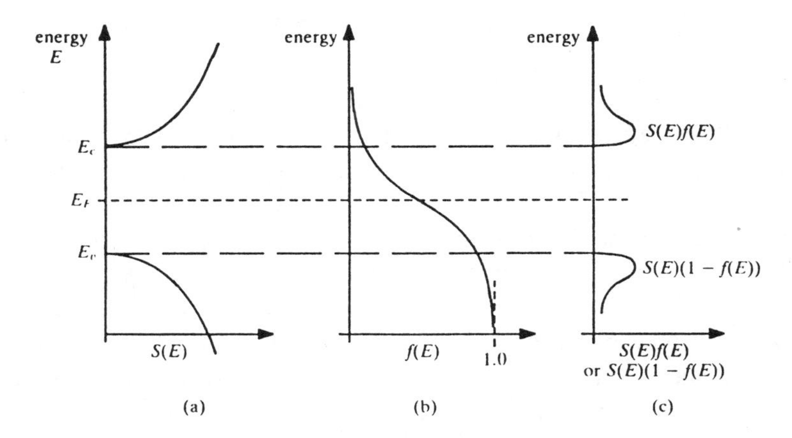

Fig. 2.15 (a) The density of states $S(E)$ (b) the Fermi function $f(E)$ (c) the electron concentration $dn(E)/dE$, all plotted against energy E. The density of states in the valence band and $dp(E)/dE$ are also shown in (a) and (c) respectively.

*See footnote page 11 for the meaning of effective mass.

Though the integration of eqn. (2.12) is possible as it stands, we can gain more insight by simplifying it with the assumption† that $\exp(E - E_F/kT) \gg 1$. Thus we neglect unity in the denominator and obtain

$$n = A\int_{E_c}^{\infty} (E - E_c)^{1/2} \exp -\left(\frac{E - E_F}{kT}\right) dE \qquad (2.13)$$

Now substitute $x = (E - E_c)/kT$ with the result

$$n = A(kT)^{3/2} \exp -\frac{(E_c - E_F)}{kT} \int_0^{\infty} x^{1/2} \exp(-x)dx \qquad (2.14)$$

The integral is not elementary, but can be shown to equal $\sqrt{\pi/2}$ yielding finally:

$$n = \frac{\sqrt{\pi}}{2} A(kT)^{3/2} \exp -\left(\frac{E_c - E_F}{kT}\right)$$

Since A is independent of temperature T, the exponential dominates the temperature variation of n. It is convenient to write

$$n = N_c \exp -\left(\frac{E_c - E_F}{kT}\right)$$

where $N_c = \left(\sqrt{\pi/2}\right)A(kT)^{3/2}$ has the value 2×10^{25} m^{-3} in silicon at 300 K. This finally justifies eqn. (2.3), and although N_c is weakly dependent on temperature, its variation can be neglected for small excursions about room temperature.

The functions $S(E)$, $f(E)$ and their product which, according to eqn. (2.10), equals $dn(E)/dE$, are all sketched in Fig. 2.15(a), (b) and (c) respectively. Note how the electron distribution is bunched close to E_c.

The most probable energy — at the peak of the distribution — can be found by equating the differential of eqn. (2.10) to zero, and is $kT/2$. The average kinetic energy, $\langle E - E_c \rangle$, can be shown using eqn. (2.10) to equal $3kT/2$, exactly as was assumed in Chapter 1, and this result helps to justify our use there of classical kinetic theory to discuss the motion of electrons.

---

†The error involved in making this assumption is less than 10 per cent as long as $\exp(E/kT) \leq 0.1$, which implies $E_c - E_F \geq 2.3kT$, about 0.06 eV at 300 K.

Now turn to the electrons and holes in the valence band. In the valence band, the general expression (2.10) still applies, but the density of states $S(E)$ is not the same as that in the conduction band. Here, the number of states is zero at E_v, and rises below E_v in the manner illustrated in Fig. 2.15(a) — it is the mirror image of the conduction band's density of states. Indeed it can be shown to obey the equation

$$S(E) = B(E_v - E)^{1/2} \tag{2.15}$$

where B depends on the effective mass of holes, m_h, and Planck's constant. The concentration of holes between energies E and $E + dE$ in the valence band is just this density of available states, multiplied by the probability $(1 - f(E))$ that they are unoccupied, which in section 2.7 was shown to be approximately*

$$\exp -\frac{E_F - E}{kT}$$

Hence

$$\begin{aligned} dp(E) &= S(E)(1 - f(E))dE \\ &\simeq B(E_v - E)^{1/2} \exp -\left(\frac{E_F - E}{kT}\right)dE \end{aligned} \tag{2.16}$$

The similarity of this equation to the corresponding one for electrons (the integrand of eqn. (2.13)) is obvious. On integrating eqn. (2.16) we make a slightly different substitution to that used in eqn. (2.13), viz. $x = (E_v - E)kT$, and obtain

$$p = B(kT)^{3/2} \exp -\left(\frac{E_F - E_v}{kT}\right)\int_0^\infty x^{1/2} \exp(-x)dx$$

The integral is identical to that in eqn. (2.14), so that

$$p = N_v \exp -\left(\frac{E_F - E_v}{kT}\right)$$

where $N_v = (\sqrt{\pi}/2)B(kT)^{3/2}$, and differs from N_c by little, since A and B are similar in magnitude. Note how these holes are distributed across the energies available in the valence band. Equation (2.16) gives the distribution,

*The error is less than 10 per cent provided that $E_F - E_v \geq 2.3kT$.

and dp/dE is plotted in Fig. 2.15(c) on the same graph as dn/dE. As before, the peak occurs at an energy $kT/2$ below E_v, and the *average* value of $(E_v - E)$ turns out to be $3kT/2$.

These results are almost, but not quite, exact. In practice, some second-order effects conspire to complicate the state of affairs. This does not, however, invalidate the simple models used here.

PROBLEMS

2.1 Contrast the energy level diagrams of a hydrogen atom and a metallic solid. Give approximate values for the separation between energy levels in each case.

2.2 State the principle that prevents all electrons in a solid from occupying the lowest available energy level.

2.3 Draw the energy level diagram of an insulator, and use it to explain why electrons in the insulator are unable to conduct a current.

2.4 Name the two most important energy bands in a semiconductor, and indicate on a diagram the distribution of electrons within each.

2.5 What does the Fermi function represent? Define the term *Fermi level*, and sketch the shape of the Fermi function for two different temperatures.

2.6 Explain qualitatively why it is that most of the holes in a semiconductor are spread across a narrow range of energies very close to the top of a band.

2.7 With the aid of an energy band diagram, illustrate the energy changes involved in
(i) direct recombination of electrons and holes (ii) absorption of light (iii) ionization of acceptor (iv) drift of electrons.

2.8 Draw an energy level diagram for silicon doped with *equal* concentrations of donors and acceptors having identical ionization energies of 0.05 eV. Sketch alongside a graph of the Fermi function in the correct relative position and deduce the fraction of donor and acceptor levels occupied by electrons at 300 K. Explain why the conductivity is the same as in pure undoped silicon.

2.9 Use eqn. (2.14) to show that dp/dE has a maximum in the valence band at $E_v - E = kT/2$, and interpret the meaning of this result.

2.10 What approximations are made in using the expressions
$n = N_c \exp - (E_c - E_F)/kT$ and $p = N_v \exp - (E_F - E_v)/kT$?

2.11 Calculate the approximate intrinsic electron concentration n_i in both silicon and germanium at a temperature of 100°C. Assume that N_c and N_v are independent of temperature, and are equal to 8×10^{24} in germanium.

Why is the percentage increase between 300 K and 393 K so much larger in silicon than it is in germanium?

2.12 Calculate an approximate figure for n_i in diamond ($E_g = 5.3$ eV), assuming that N_c and N_v are similar to their values in Si and Ge. Hence show that the measured conductivity of about 10^{-12} S/m at 300 K is unlikely to be due to the due to the simple motion of electron and holes.

2.13 The energy gap of SiO_2 is about 8 eV. At what wavelength should it begin to absorb electromagnetic waves strongly?

2.14 When electrons and holes recombine in GaAs, the energy released is emitted as photons. What is the *maximum* wavelength emitted, given that the energy gap is 1.43 eV?

Use the result of problem 2.9 to estimate the wavelength of the most intense radiation emitted at 300 K assuming that the probability of recombination is the same for all electrons and holes.

2.15 Find an approximate upper doping limit in silicon, at $T = 150$ K, beyond which donors or acceptors are less than 90 per cent ionized. Comment on the difference between this figure and the figure quoted in the text for $T = 300$ K.

Chapter 3

p–n Junction Diodes

The rectifying action of a junction between *p*-type and *n*-type semiconductors is of universal utility in integrated circuits. The insulating property of a reverse-biased junction is the means by which components are isolated from one another. The prime aim of this chapter is to derive an equation to describe the *I–V* characteristic of a *p–n* junction. It will alse explore briefly four other applications: the varactor diode (a voltage-controlled variable capacitor), the light-emitting diode, the photodiode and the Schottky diode.

In addition to deriving the current-voltage characteristic, we will also present another frequently-used type of model for an electronic device: the *small-signal equivalent circuit.* This represents the behaviour of the device when a small alternating voltage δV is applied to the terminals. Since the current and voltage in a diode are not proportional to one another (we say that the *I–V* relation is *non-linear*), then V/I and $\delta V/\delta I$ cannot be equal to one another, and the small-signal resistance — which is a component of the equivalent circuit — has to be calculated from the equation relating I and V. But first we must consider the potential distribution within a *p–n* junction having no bias voltage applied across it.

PANEL 3.1

p–n Diodes: fabrication and impurity profiles

In integrated circuits, the process of diffusing, say, *p*-type impurities at a temperature of about 1000°C into an *n*-type wafer from one surface as at (A) produces an impurity profile which is rather gradual (B). Abrupt changes in doping can be obtained by making the junction at a lower temperature so that diffusion is slowed. A small drop of liquid metal dopant (e.g. indium, melting point 156°C) placed on the wafer surface (C) dissolves germanium up to the limit of its solubility in indium. As the wafer cools, this molten alloy solidifies in the same crystal form and orientation as the wafer. The doping profile is given at (D).

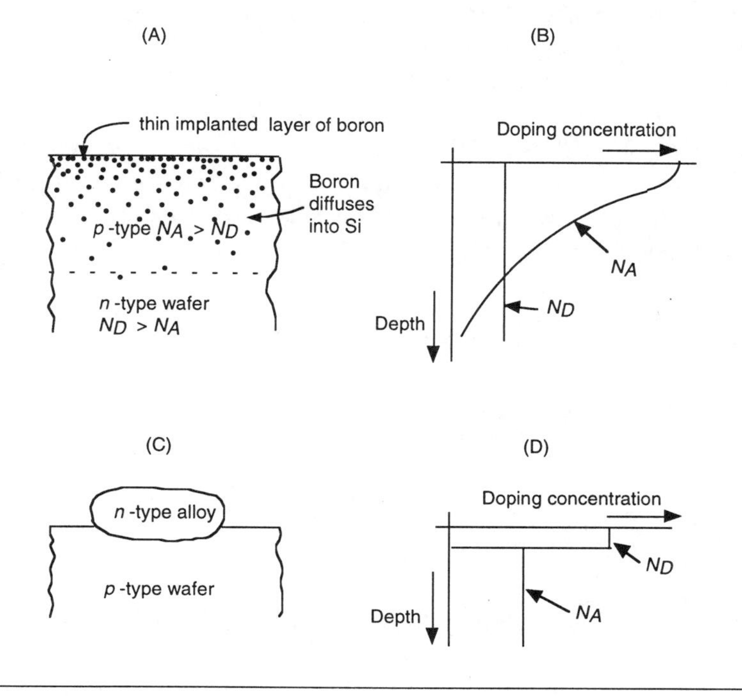

3.1 The junction in equilibrium (no bias)

A rectifying *p–n* junction cannot be made simply by joining two pieces of semiconductor, for the resulting join is too imperfect for conduction across it to occur in the manner described below. The crystal lattice must be perfectly continuous across the 'junction' — only the doping may differ on either side — so that junctions are most commonly made by diffusion, e.g. of a *p*-type impurity, such as boron, into an *n*-type semiconductor.

For simplicity of discussion, we assume an abrupt change in doping at the junction as plotted in Fig. 3.1(a). Although this is not the case in integrated circuits (see Panel 3.1) the differences in behaviour are small, and will be mentioned later.

Now the current carriers are mobile, and will diffuse down a concentration gradient. Their concentrations cannot therefore change abruptly across the junction as illustrated in Fig. 3.1(b), where they are shown together with their parent donors and acceptors which are, of course, immobile. However, a gradient of concentration is unavoidable, so diffusion across the junction occurs continuously. It can only be balanced by a drift current in the other direction — but there is no obvious electric field to maintain such a drift. We shall now show how a field arises from the *built-in potential difference* that exists between any two dissimilar materials, and which originates in a transfer of charge between the two. The linkage between these processes is illustrated below.

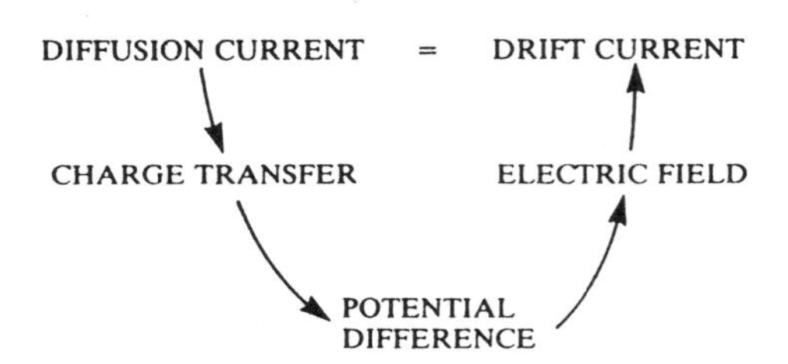

The detail of these relationships will be explained with the aid of Fig. 3.1(c)–(f).

The easiest way to appreciate the final result is by considering how it might develop if the two halves of the junction were abruptly placed in contact with one another (a physical impossibility without trapping

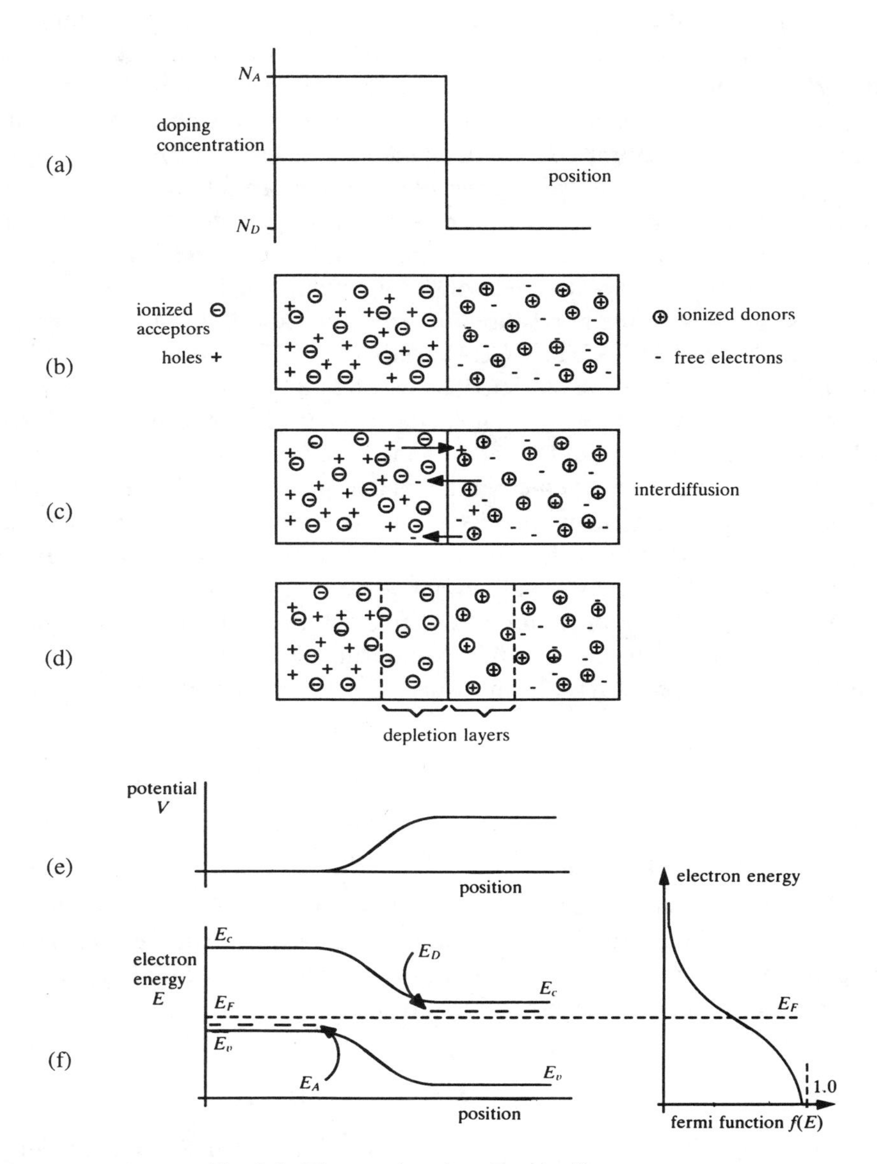

Fig. 3.1 The *p–n* junction. For details see text.

contamination at the interface, so this is a 'thought experiment'). Initially, interdiffusion of carriers leads to recombination as they mingle close on either side of the junction (Fig. 3.1, diagram (c)). Hence near the junction their numbers fall dramatically. But, in removing negatively charged electrons from the originally neutral n-side of the junction, the latter becomes *positively* charged in the thin layer where the electron concentration is depressed. These positive charges are actually the single excess fixed charges on each of the nuclei of the ionized donor atoms. Similarly, a negatively charged region develops on the p-type side of the junction, consisting of the excess electrons captured in the bonds holding each acceptor atom in place. These regions are termed DEPLETION LAYERS. The final result is illustrated in Fig. 3.1(d), where we have assumed, as we shall later show, that the mobile carriers are almost entirely absent in the depletion layers, up to a fairly well-defined distance on either side of the junction.

It is the charges in the depletion layers that give rise to the electric field in the junction. The field is directed *from* the positive charges *toward* the negative ones, causing a drift current of holes back toward the p-side from which they are diffusing — and similarly with electrons.

Once the balance of drift and diffusion is achieved, recombination plays no role in maintaining the dynamic equilibrium, for the electron and hole currents given by eqns. (1.22) and (1.23) are each equal to zero:

$$J_n = ne\mu_e E + D_e e\frac{dn}{dx} = 0$$

$$J_p = pe\mu_e E - D_h e\frac{dp}{dx} = 0$$

We shall look at the solutions of these equations later. For the present we shall construct the energy level diagram for the junction. The field E, which varies with x, is associated with a built-in electrostatic potential difference, V_0 between the p- and the n-regions, given by

$$V_0 = -\int_{\text{p-side}}^{\text{n-side}} E\,dx$$

as illustrated in Fig. 3.1(e). The potential rises steadily across the depletion layers, so that the potential energy of a free electron is *higher* to the left (p-side) of the junction than to the right (n-side). This enables us to re-draw the electron energy diagrams of p-type and n-type regions as in Fig. 3.1(f).

The most significant feature of this diagram is that the Fermi energy is drawn at the *same level* right across the junction. This is because, in equilibrium, the probability of an electron having a given energy must be the same wherever it may be in the one piece of material comprising the junction. Otherwise, the Fermi distribution could not describe equilibrium.

The potential energy difference $E_{cp} - E_{cn}$ between free electrons in the conduction band on either side of the junction is just equal to eV_0, so we can now relate V_0 to the electron concentrations on either side, because, from eqns. (2.3) and (2.4)

$$n_n = N_c \exp -(E_{cn} - E_F)/kT$$

$$\text{and} \quad n_p = N_c \exp -(E_{cp} - E_F)/kT$$

Thus by division,

$$\frac{n_n}{n_p} = \exp\left(\frac{E_{cp} - E_{cn}}{kT}\right) = \exp\frac{eV_0}{kT} \tag{3.1}$$

Taking logarithms gives the result:

$$V_0 = \frac{kT}{e}\ln\left(\frac{n_n}{n_p}\right) \tag{3.2}$$

By putting n_n and n_p in terms of doping densities N_A and N_D:

$$V_0 = \frac{kT}{e}\ln\left(\frac{N_D N_A}{n_i^2}\right)$$

For the moment, notice that V_0 depends only logarithmically on N_A and N_D and so varies little from diode to diode. Typically, in silicon we may calculate V_0 for $N_A = 10^{22}$, $N_D = 10^{24}$:

$$V_0 = 0.82\ \text{V at } 300\ \text{K in Si}$$

Because n_i^2 is much larger in Ge, V_0 is smaller for a Ge junction (see problem 3.9).

This potential difference, like all contact potentials, cannot be measured directly with an ordinary voltmeter, because there is no source of energy to drive even a tiny current through the instrument.

3.2 Current balance in the junction

We pointed out in the previous section that the electron and hole currents are individually zero. Thus, using eqn. (1.22) we have

$$ne\mu_e E = -D_e e \frac{dn}{dx} \tag{3.3}$$

Putting $E = -dV/dx$ and rearranging, we have:

$$\frac{dV}{dx} = \frac{D_e}{\mu_e}\left(\frac{1}{n}\frac{dn}{dx}\right)$$

This equation relates the potential variations to the electron concentration changes. If we now integrate this equation with respect to x between the points $x = -\infty$ and $x = \infty$, we have

$$\int_{-\infty}^{+\infty} \frac{dV}{dx}\,dx = \frac{D_e}{\mu_e}\int_{-\infty}^{+\infty} \frac{1}{n}\frac{dn}{dx}\,dx$$

Now change the variable of integration to V on the left, and to n on the right hand side.

The limits of integration for V can be made just 0 and V_0 (the potential can be made zero at any arbitrary point), while for n they are n_p and n_n. Performing the integration gives

$$V_0 = \frac{D_e}{\mu_e}\ln\left(\frac{n_n}{n_p}\right) \tag{3.4}$$

This looks very like eqn. (3.2) derived above; indeed use of Einstein's relation $D/\mu = kT/e$ shows that they are identical.

This is a comforting result, as the 'proof' of eqn. (3.4) is dubious, since the occurrence of diffusion within the junction region is not obvious: the mean free path is normally no smaller than the thickness of the depletion layer. The importance of eqn. (3.4) is that the same equations are needed for discussing the junction when a bias voltage is applied across it.

3.3 The potential barrier and the effects of bias

We can regard the potential difference V_0 as a 'barrier' to flow of majority carriers across the junction: in Fig. 3.1(f) the shape of the conduction band edge resembles a 'hill' which the electrons in the *n*-type region are unable to climb. Remembering that holes naturally 'float' to the top of the valence band, the shape of its edge also represents a 'hill' in their antipodean world, preventing them from leaving the *p*-region with ease.

We now show that this barrier height can be varied by applying an external voltage across the junction, making the flow of the majority carriers across it easier or more difficult according to the sign of the voltage: this is the origin of the rectifying action.

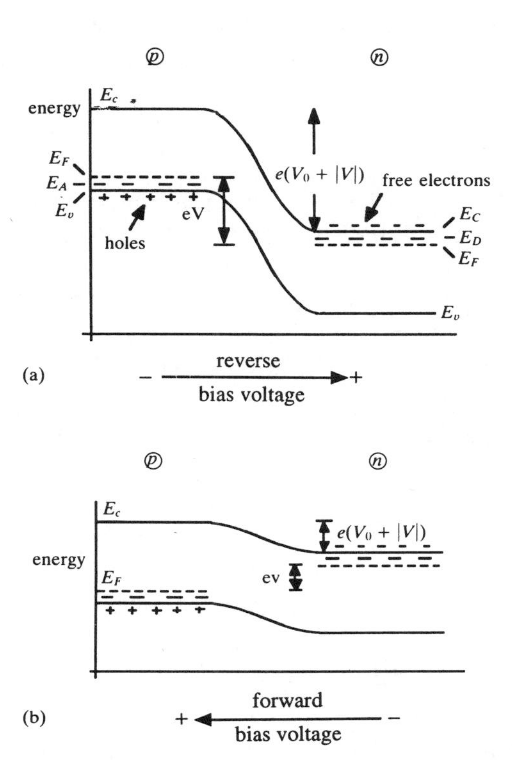

Fig. 3.2 Energy level diagrams for a *p–n* junction under (a) reverse bias (b) forward bias.

We have seen that the whole of the 'built-in' potential difference V_0 is dropped across the depletion layers, where there are extremely few carriers, so that these layers have a high resistance compared to the neutral regions on either side. Thus when we apply bias, the added voltage drop V does not occur in the regions where carriers are plentiful, but within the resistive depletion layers. The new energy level diagrams are shown in Fig. 3.2; the potential energy barrier is higher when the *p*-type side is made more negative (REVERSE BIAS) and lower when it is more positive (FORWARD BIAS). Note the energy difference eV between the Fermi levels on either side. If we assign the voltage V a positive sign when in forward bias and a negative sign when in a reverse bias, the height of the potential barrier in either case is $(V_0 - V)e$.

We now consider how to calculate the current densities J_n and J_p. The first assumption we shall make is that there is negligible recombination anywhere — i.e. the whole diode is short compared to a diffusion length. The current density J_n is determined by the usual equation:

$$J_n = -ne\mu_e \frac{dV}{dx} + D_e e \frac{dn}{dx} \tag{3.5}$$

but recognizing that n no longer has its equilibrium values everywhere, and the new values must be found. We can find these values at the edges of the depletion layer by using a method like that in the previous section. But first we must make a simplifying approximation in eqn. (3.5), which may seem drastic. We can show that J_n is usually so much smaller than either of the other terms in eqn. (3.5), that it can be neglected compared to them.

The diffusion term can be roughly estimated by borrowing the result of a later calculation, that the width of a depletion layer is typically 0.1 μm. Then the concentration gradient dn/dx is of the order of $n/W \sim 10^{24}/10^{-7} = 10^{31}$ m^{-4}, say. Thus $D_e e(dn/dx) = \mu_e kT(dn/dx) \sim 10^{10}$ A/m^2, compared to which the largest likely current density of $\sim 10^7$ A/m^2 is indeed negligible. It is now possible to see that eqn. (3.3) still holds good, to an accuracy better than 0.1 per cent. Hence we may integrate it as before:

$$\int_0^{V_0 - V} dV = \frac{D_e}{\mu_e} \int_{n_p'}^{n_n} \frac{dn}{n}$$

Note carefully the new limits in the integral on the right: we know the potential difference $(V_0 - V)$, but not yet the value n_p' of the electron concentration at the depletion layer edge in the p-type material. The upper limit remains n_n, as will shortly be explained. Performing the integration yields:

$$V_0 - V = \frac{D_e}{\mu_e} \ln\left(\frac{n_n}{n'_p}\right)$$

Or, using Einstein's relation,

$$V_0 - V = \frac{kT}{e} \ln\left(\frac{n_n}{n'_p}\right) \tag{3.6}$$

Why is n_p increased to n'_p, while n_n is unaltered? The answer lies in the magnitudes of n_n (a majority carrier concentration) and n_p (a minority carrier concentration) relative to the changes in concentration. Typically, the changes in electron concentration might be around 10^{19} m^{-3}, much *smaller* than n_n (which may be around 10^{22} m^{-3}), but very much *larger* than n_p (which is typically 10^{10} m^{-3}). So the concentration of electrons on the n-type side is barely altered from its equilibrium value n_n,* while n'_p undergoes a manifold increase over its equilibrium value n_p. By rearranging eqn. (3.6) an expression for n'_p is found in terms of V:

$$n'_p = n_n \exp \frac{e(V - V_0)}{kT}$$

But eqn. (3.2) can similarly be arranged to read:

$$n_p = n_n \exp \frac{-eV_0}{kT}$$

Whereupon we may divide these two equations, with the final result

$$\frac{n'_p}{n_p} = \exp \frac{eV}{kT} \tag{3.7}$$

---

*Surprisingly, n_n actually *increases* very slightly, in order to balance the injection of positively-charged holes into the n-type region, so maintaining local electrical neutrality. This small change can be neglected in these calculations.

This show that n_p' is indeed very much larger than n_p for voltages greater than a few times kT/e, for example, 0.1 V or more.

In an exactly similar way we can show that the hole concentration p_n' on the n side of the junction is raised by the same factor, i.e.

$$\frac{p_n'}{p_n} = \exp\frac{eV}{kT} \tag{3.8}$$

The circuit designer's rule of thumb for the 'on' state voltage of a diode is 0.6–0.7 V, and at this voltage the ratio n_p'/n_p or p_n'/p_n is 10^{10}–10^{11}. With this result we can fill in some typical numbers on the concentration graphs in Fig. 3.3 which are not drawn to scale.

Assuming the junction dopings are $N_D = 10^{23}$ m^{-3}, $N_A = 2 \times 10^{24}$ m^{-3}, then $n_p = n_i^2/N_A = 10^8$ m^{-3}. With a bias $V = 0.6$ V across the junction, eqn. (3.7) gives $n_p' = 10^{18}$ m^{-3}. We can now see more clearly that the majority carrier concentration at the depletion layer edge changes very little. For to maintain charge neutrality on the p-side when n_p' rises to 10^{18} m^{-3}, the concentration p_p, originally 10^{24} m^{-3}, must rise by 10^{18} m^{-3} — a negligible change.

Readers can calculate the values of p_n, p_n' and the change in n_n for themselves in a similar way.

The excess holes injected into the neutral n-type region, where $dV/dx = 0$, can only move onward by diffusion. Hence the concentration gradient in the neutral region can be used to work out the hole current density. Having assumed that the diode is 'short' enough for recombination to be negligible, *the hole current in the n-type region must equal the hole current on the p-side of the junction.* On the n-side (see Fig. 3.3) the hole concentration falls linearly (in the absence of recombination) towards its equilibrium value at the contact point, a distance L_n from the junction. (See Panel 3.2 for a discussion of contacts.)

The hole current density is thus

$$J_p = -D_h e\frac{dp}{dx} = D_h e\left(\frac{p_n' - p_n}{L_n}\right)$$

Note that the negative value of dp/dx ensures that J_p has a positive sign, i.e. it flows in the $+x$ direction.

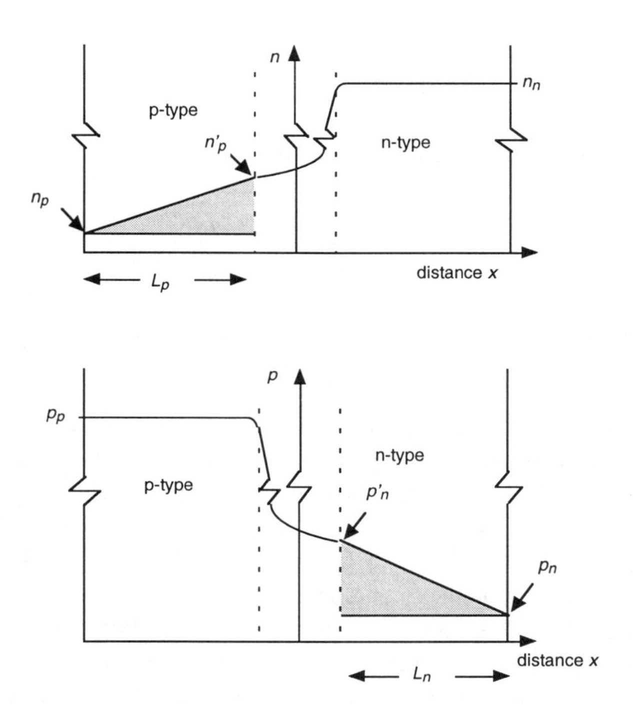

Fig. 3.3 Electron and hole concentrations across a forward-biased *p–n* junction. Note the change in scale on the vertical axis.

Now using eqn. (3.8) for p_n', we can express J_p in terms of the applied voltage:

$$J_p = \frac{D_h e p_n}{L_n}\left[\exp\left(\frac{eV}{kT}\right) - 1\right] \tag{3.9}$$

The electron current density can, of course, be derived in exactly the same way:

$$J_n = D_e e \frac{dn}{dx} = D_e e\left(\frac{n_p' - n_p}{L_p}\right) = \frac{D_e e n_p}{L_p}\left[\exp\left(\frac{eV}{kT}\right) - 1\right] \tag{3.10}$$

Adding together the current densities given by these two equations leads us to the universal diode current equation

$$I = I_s\left[\exp\left(\frac{eV}{kT}\right) - 1\right] \tag{3.11}$$

Where we see that for a junction with area A:

$$I_s = A\left(\frac{D_e e n_p}{L_p} + \frac{D_h e p_n}{L_n}\right)$$

PANEL 3.2

Contact metals must be chosen to avoid making a rectifying junction. Given a suitable metal, it can supply to, or receive from, the semiconductor any number of electrons, preventing any attempt to disturb the equilibrium concentrations of carriers at the contact surface. This justifies the assumption in Fig. 3.3 that $p = p_n$ at the contact.

Aluminium is a metal often chosen for contact to p-type Si, but on n-type Si it forms a rectifying contact. This is because a potential barrier is formed at the junction. By doping the n-type Si very heavily the potential barrier can be made very thin indeed. Electrons are then able to 'tunnel' through the barrier due to their wavelike properties, and the contact is ohmic rather than rectifying. Hence Al contacts to n-type wafers are made only where an n^+ diffusion (a term for heavy doping) has been carried out.

This peculiar behaviour of electrons is analogous to the way light can 'leak' through a micrometre thick film of air separating two glass surfaces, even when the angle of incidence in the glass is greater than the critical angle.

3.4 Reverse bias

In the derivation of the current equation, we made no assumptions that apply exclusively to forward bias: remember that the potential difference across the depletion region is $V_0 - V$ also in reverse bias, if V is made negative when the bias voltage is reversed. Thus all the equations derived above can be used if this convention is observed.

Hence eqn. (3.7) tells us that if the reverse voltage is more than a few times kT/e, then $\exp eV/kT \ll 1$, and so $n'_p \ll n_p$. For example, if $V = -1$ V, $n'_p/n_p \simeq 10^{-17}$, according to eqn. (3.7). So we can draw the electron distribution across the diode as shown in Fig. 3.4. Instead of electrons (and holes) being injected into either side across the barrier, minority carriers are 'sucked' out by the high electric field strength in the junction (shown in Fig. 3.2(a)), as soon as they diffuse to the edge of the depletion layer.

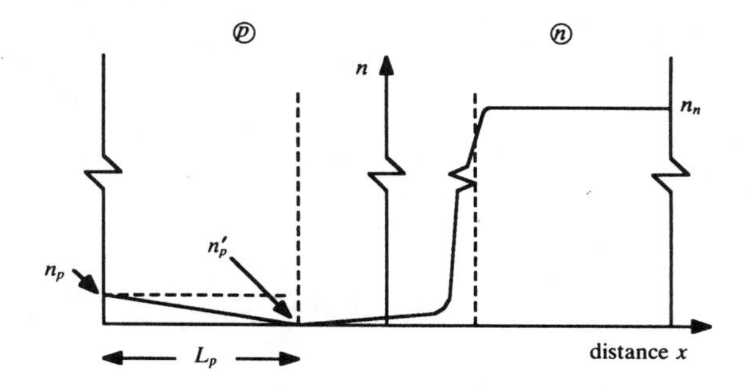

Fig. 3.4 Electron concentration across a reverse-biased junction. Note the change on scale on the vertical axis.

The gradient of minority carriers in the neutral regions is now reversed, and they diffuse *towards* the junction. The current they carry can be calculated from the gradient in Fig. 3.4:

$$J_n = D_e e \frac{dn}{dx} = D_e e \left(\frac{n_p - n'_p}{L_p} \right) = \frac{D_e e n_p}{L_p}$$

Similarly

$$J_p = \frac{D_h e p_n}{L_n}$$

These are identical to the currents calculated by letting $V \to -\infty$ in eqns. (3.9) and (3.10).

3.5 The current–voltage characteristic

The I–V characteristic predicted by the above theory in eqn. (3.11) is illustrated in Figs. 3.5(a) and (b) — in these universal curves I and V are plotted as multiples of I_s and kT/e respectively. Note that the reverse current saturates for reverse voltages of more than about $-3(kT/e)$ (because e^{-3} is only about 0.05). In the forward direction, the current rises very rapidly but smoothly: there is no real 'threshold' as sometimes assumed in books on circuit theory. However, as in Fig. 3.5(b) shows, in strong forward bias, the diode voltage varies rather little over a wide current range. For example, a diode designed to carry a current I at, say, 0.6 V (i.e. $eV/kT = 24$) will carry

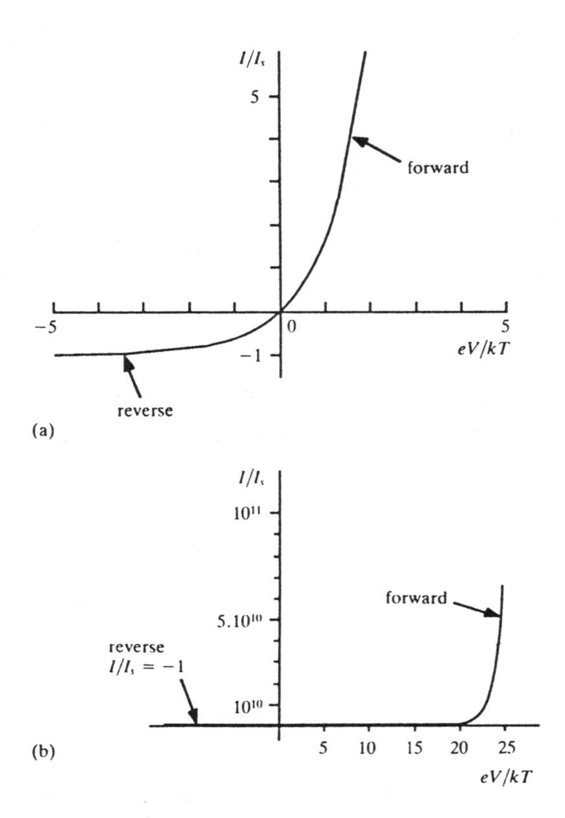

Fig. 3.5 The current–voltage characteristic on two different scales.

a current of only 0.001 I at 0.427 V ($eV/kT = 17.1$). For practical purposes the diode is 'off' below this voltage.

The difference in the 'on' voltages of Ge and Si diodes also needs explanation. Note that the 'on' current for a Ge diode with the same applied voltage as an Si diode (e.g. 0.7 V) is the same multiple of its reverse saturation current I_s. Since I_s is proportional to the *minority* carrier density, which in turn depends on n_i^2, the reverse saturation current of a Ge diode — and hence the forward current — is about 10^8 times that of an Si diode with the same doping levels. Thus a Ge diode 'turns on' at a lower voltage — about 0.25 V is typical. Note that, to make an Si diode turn on at this voltage would need either a doping level about 10^8 times smaller, or an area 10^8 times bigger: neither is a practical proposition.

Another interesting implication of these results is, that any diode with a low 'turn on' voltage must also have a rather high reverse saturation current; for example Ge should be about 10^8 times worse than Si in this respect, though it is not quite so much worse in practice, as we shall now explain.

3.6 Characteristics of a practical diode

Real diodes differ from the above ideal behaviour, because we have deliberately ignored three factors:

(a) Generation and recombination of carriers within the narrow depletion region.

Carriers generated thermally within the depletion region are subjected to the large electric field there, and so that any holes generated are driven to the *p*-side of the junction, and electrons to the *n*-side, before they have a chance to recombine. When no bias is applied, the current which results is balanced by the recombination in the depletion region of carriers entering it by thermal diffusion. Under a voltage bias, these currents no longer balance, giving an extra term in the current equation. Thus the current-voltage relation in real diodes is accurately modelled by an equation having two terms:

$$I = I_s\left[\exp\frac{eV}{kT} - 1\right] + I_{Ro}\left[\exp\frac{eV}{2kT} - 1\right] \tag{3.12}$$

The first term, which is normally dominant in forward bias, is exactly as given in eqn. (3.11).

The second term, due to carrier generation and recombination in the depletion region, has an extra factor of 2 in the denominator of the exponent. Hence it varies roughly as the square root of the first term. Thus at small forward currents, and in reverse bias (i.e. $V < 0$), the second term becomes an important factor. The pre-factor I_{Ro} is proportional to the volume in which this generation and recombination is occuring, i.e. the volume of the depletion regions. Hence I_{Ro} increases with the increase in depletion-layer width which occurs in reverse bias (see section 3.7), and the reverse current is thus prevented from saturating in a real diode.

At sufficiently large forward currents the term containing I_{Ro} becomes negligible, and eqn. (3.11) is a valid approximation.

Another contribution to the observed reverse current may come from the finite resistivity of the insulators used to package the diode, and from surface leakage currents between connecting pins, if any.

(b) Voltage drops across the 'neutral' region have also been ignored. At high forward currents the diode current is limited more by the small series resistance R_s in the bulk semiconductor. As a result, the voltage across a junction carrying a high current is smaller by an amount IR_s than the total applied voltage. Thus eqn. (3.13) is a more accurate representation at high currents:

$$I = I_s\left[\exp\left(\frac{e(V - IR_s)}{kT}\right) - 1\right] \qquad (3.13)$$

(c) Beyond a few volts of reverse bias, the diode current rises dramatically, a phenomenon termed BREAKDOWN. In spite of the name, this may not cause irreversible damage to the diode if the current is limited by the circuit, to prevent overheating. Breakdown results from the extremely high electric field in the depletion layer of a reverse-biased junction.

There are two mechanisms of breakdown: which of these occurs in a particular diode depends on the width of the depletion layer. If it happens to be relatively wide, the few carriers crossing it can be accelerated by the field to quite high drift velocities. On colliding with the lattice they can give up enough of their kinetic energy to create new electron-hole pairs. This AVALANCHE process, as it is called, causes the current to increase extremely rapidly with voltage above a breakdown value, which can be as high as 1 kV in a properly designed power diode.

If the depletion layer is thin enough for the breakdown voltage to be less than about 5 V, a different process, called electron tunnelling, occurs (for details see a more advanced text). The current rises rapidly with voltage in the same way.

The fact that a diode can withstand the breakdown process is exploited in the ZENER DIODE. Because the voltage across a diode in breakdown is almost independent of the current, the diode can be used as a stable voltage reference. Zener diodes are available with breakdown voltages from a few volts up to about 270 V.

3.7 The depletion layer width

In section 3.3 we were able to derive the *I–V* characteristic without considering the detail of the charge and voltage distribution across the depletion region. We shall now deduce the width of this layer, first in the equilibrium case, when no bias is applied.

The charge distribution depends upon how many carriers are left at each point. Since in equilibrium $n = N_c \exp(E_c - E_F)/kT$ we can in principle find n from a knowledge of $E_c - E_F$. But, looking back at the energy level diagram of the junction, Fig. 3.1(f), we see that $E_c - E_F$ is noticeably larger than its bulk value almost all the way across the depletion region. Because the exponential dependence on $E_c - E_F$ is so strong, the electron concentration n falls like a cliff, from its bulk value on the edge of the depletion layer to practically zero just inside it, as shown in Fig. 3.6(a). The charge concentration inside the depletion layer thus equals eN_D (or $-eN_A$ on the *p*-side) and is shown plotted in Fig. 3.6(a).

Since this distribution looks so like a rectangle, we shall approximate it by one. In other words, we assume that the charge density equals eN_D from $x = 0$ to the junction at $x = W_n$, and and is $-eN_A$ beyond the junction, to a point at which the *p*-side depletion layer is assumed to end abruptly, at $x = W_n + W_p$. The voltage drop across this charge distribution can then be calculated by applying Gauss' law to the cylinder shown in Fig. 3.6(c), whose left-hand face lies at the depletion layer edge.

The electric flux leaving the cylinder is zero on the left-hand face because all the flux lines beginning on positive charges must end on negative charges, and so must be directed to the right. By symmetry, the flux lines must be

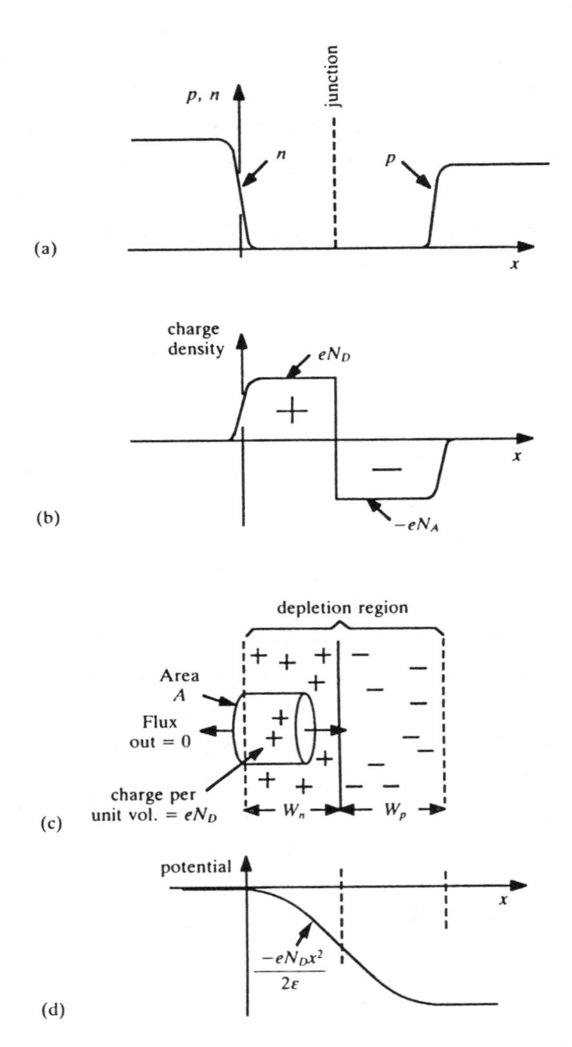

Fig. 3.6 (a) Carrier concentration and (b) resultant net charge distribution across the depletion layer (c) calculation of the field strength in the charged region (d) calculated potential variation across the junction.

parallel to the curved surfaces of the cylinder, so no flux passes through that, either. The total flux leaving the cylinder thus equals that through the right-hand face, whose area is A. This flux equals $\varepsilon E(x)A$, where $E(x)$ is the electric field strength at the distance x from the depletion layer edge and ε is the permittivity of silicon. By Gauss' law, the flux out of the cylinder equals the charge enclosed in it, which is just charge density × volume, i.e. eN_DAx.

Thus

$$\varepsilon E(x)A = eN_DAx$$

i.e.

$$E(x) = -\frac{dV}{dx} = \frac{eN_Dx}{\varepsilon}$$

Now integrate this equation to find $V(x)$ at the point x

$$V(x) - V(0) = -\int_0^x \frac{eN_Dx}{\varepsilon}dx = -\frac{eN_Dx^2}{2\varepsilon} \tag{3.14}$$

So $V(x)$ varies parabolically with x up to the centre of the junction, as plotted in Fig. 3.6(d). The potential difference between $x = 0$ and $x = W_n$ is therefore equal to $eN_DW_n^2/2\varepsilon$. An exactly similar argument leads to an equivalent expression for the potential difference across the depletion layer on the p-side of the junction; the result is $eN_AW_p^2/2\varepsilon$.

Now the sum of these two potential differences must be equal to the total potential drop across the p–n junction, which is just the built-in potential V_o, less any forward bias voltage V which is externally applied. Equating these potential differences thus gives

$$V_0 - V = \frac{eN_DW_n^2}{2\varepsilon} + \frac{eN_AW_p^2}{2\varepsilon} \tag{3.15}$$

To find W_n and W_p separately from eqn. (3.15), we need an additional equation which expresses the fact that the crystal as a whole is neutral, so that the charges on the two sides of the junction are equal, thus

$$W_pN_A = W_nN_D \tag{3.16}$$

Substituting $W_p = W_n N_D / N_A$ into eqn. (3.15) and solving for W_n gives

$$W_n^2 = \frac{2\varepsilon}{e} \cdot \frac{N_A}{\left[N_A N_D + N_D^2\right]} (V_0 - V) \tag{3.17}$$

This result can be expressed in terms of the width W_{no} of the depletion layer at zero bias:

$$W_n^2 = W_{no}^2 \left(1 - \frac{V}{V_0}\right) \tag{3.18}$$

where

$$W_{no}^2 = \frac{2\varepsilon V_0}{e} \frac{N_A}{(N_A N_D + N_D^2)}$$

An expression for W_{po} can be similarly found to be

$$W_{po}^2 = \frac{2\varepsilon V_0}{e} \frac{N_D}{(N_A N_D + N_A^2)}$$

Putting $N_D = N_A = 10^{23}$, $V_0 = 0.82$ V and $\varepsilon = 11.7\varepsilon_0$, we find $W_{po} = W_{no} =$ 80 nm. Notice, too, that W_n and W_p are both voltage-dependent, becoming smaller under forward bias and larger under reverse bias. Further, that higher doping on one side of the junction narrows the depletion layer on the same side, but broadens that on the other side. This a feature which finds use in transistor design (see section 5.10).

One consequence of the doping dependence of W_n and W_p is that heavily doped junctions break down more readily in reverse bias, because the voltage is applied across a thinner depletion region, creating a larger electric field strength there. Thus the reverse breakdown voltage of a typical light-emitting diode (LED) is about 1 V — the need for heavy doping is explained in section 3.11. By contrast, the use of low doping levels makes it possible for rectifying diodes to be made with breakdown voltages above 1 kV.

The above equations in this section, derived for an abrupt doping profile, do not apply to diffused junctions, in which the doping varies approximately linearly with distance from the centre of the junction. For example, in a linear doping profile, eqn. (3.18) is replaced by the relation:

$$W_n^3 = W_{no}^3(1 - V/V_0)$$

Notice, however, that the derivation of eqn. (3.11) for the current-voltage relation did not depend on any assumptions about the shape of the doping distribution, and so that equation is valid irrespective of the doping profile.

3.8 The varactor diode

There are two capacitive effects in a *p–n* diode. The largest effect arises because excess injected minority carriers in the neutral, diffusion region cannot instantaneously be removed or increased in number. This is discussed in section 3.14. For the present note that, in reverse bias, the effect is small because the minority carrier density is much smaller than in forward bias.

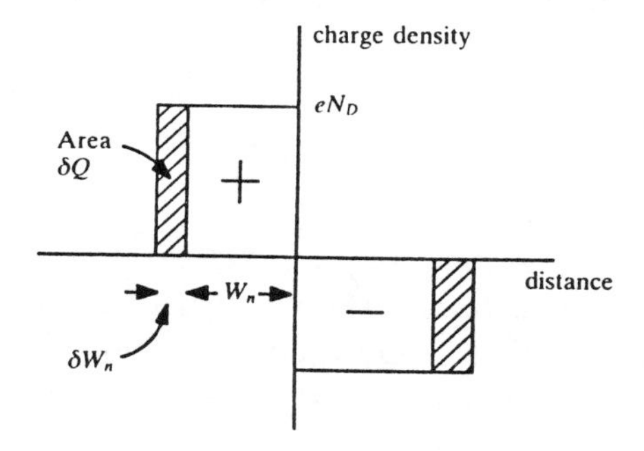

Fig. 3.7 Illustrating the variation of depletion layer charge and width with junction voltage.

The second effect dominates in reverse bias, and is due to the changes in charge stored in the depletion layers as they widen with increased reverse voltage. Figure 3.7, a re-drawn version of an earlier diagram, illustrates this: the shaded areas δQ are added to the depletion layers when the reverse bias increases by, say, δV. However, δQ is not quite proportional to δV, since eqn. (3.17) shows that $W_n \propto (V_0 - V)^{1/2}$, so that the diode is not an ideal capacitor. But a SMALL-SIGNAL CAPACITANCE per unit junction area $C_j = \delta Q/\delta V$

can be defined when $\delta V \ll V$. This is found by using Fig. 3.7 to calculate the added charge δQ from the shaded area:

$$C_j = \frac{\delta Q}{\delta V} = \frac{eN_D \delta W_n}{\delta V} \cong eN_D \frac{dW_n}{dV}$$

Now use eqn. (3.18) to find W_n:

$$W_n = W_{no}\left(\frac{1-V}{V_0}\right)^{1/2}$$

whence, by differentiating, remembering that $V < 0$ in reverse bias:

$$C_j = \frac{eN_D W_{no}}{2V_0(1-V/V_0)^{1/2}} \tag{3.19}$$

where $V_0 - V$ **increases** with reverse bias because $V < 0$. Hence the small-signal capacitance decreases roughly as the inverse square root of the reverse voltage when $V_0 \ll |V|$:

$$C_j \propto \frac{1}{|V|^{1/2}}$$

This fact can be put to good use in a variety of ways, since it represents a voltage-controlled capacitance. A diode specifically designed for such an application is called a VARACTOR diode. Remember that a diffused diode has a doping profile that is more nearly linear than abrupt (see Panel 3.1). An ideal, linear, profile can be shown to give a capacitance that varies as

$$C_j = C_{j0}/(1-V/V_0)^m \tag{3.20}$$

in which $m = 1/3$, and C_{j0} is the capacitance at zero applied voltage. A real, diffused, diode, as in an integrated circuit, has a similar capacitance-voltage relationship, but with a value of m between 1/2 and 1/3.

In practice, a measurement of the relationship between C and V can be used to determine the shape of the doping profile, and is indeed a standard method of doing so.

It is worth remembering that every diffused region in an integrated circuit has a capacitance to the substrate. This includes resistors, MOSFET sources and drains, interconnections, and particularly the large 'pads' to which wires are bonded for external connections.

3.9 Recombination and lifetime

We have avoided discussing the effects of recombination on diode behaviour, by assuming that the diode is so short that recombination can be neglected. In various devices, recombination is a significant factor, so it should be shown that it has no profound effect on the current-voltage characteristics of junction diodes. But first it will be shown how an excess of minority carriers injected across a junction are characterized by a 'lifetime', which is the average time each survives before recombining with a majority carrier.

Suppose, for the moment, that the injected concentration of minority carriers is somehow made uniform, and that no current flows, either by drift or diffusion. Then, being out of equilibrium, the concentration n of electrons decreases towards its equilibrium value, which we shall denote by n_p. The decrease occurs because the rate of recombination has risen, being proportional to the product np.

The *net* rate at which n decreases must be proportional not to n itself, but to the excess concentration $(n - n_p)$, since, when $n = n_p$, the net rate dn/dt must be zero. Thus we can assume that

$$\frac{dn}{dt} = -c(n - n_p) \tag{3.21}$$

where c is a constant.

The solution of this differential equation for n is of the form

$$(n - n_p) = A \exp - ct$$

The value of A is found by noting that, when $t = 0$, the value of $(n - n_p)$, which we write as $(n(0) - n_p)$, equals A. Putting $c = 1/\tau_e$ we thus have

$$(n - n_p) = (n(0) - n_p) \exp \frac{-t}{\tau_e} \tag{3.22}$$

where τ_e is called the LIFETIME of electrons in the p-type semiconductor and can be shown to equal the average time of survival of the excess electrons. A similar definition applies to the lifetime τ_h of holes in an n-type sample. Note that by putting $c = 1/\tau_e$ into eqn. (3.21) it becomes

$$\frac{dn}{dt} = -\left(\frac{n - n_p}{\tau_e}\right) \tag{3.23}$$

while a similar equation applies to holes:

$$\frac{dp}{dt} = -\left(\frac{p - p_n}{\tau_h}\right) \tag{3.23a}$$

Mechanisms of recombination are varied, the dominant factors in silicon and germanium being the 'catalytic' effect of defects, especially at surfaces and impurity atoms (see Panel 3.3). Hence both τ_h and τ_e vary widely, typically over the range 10^{-8} s to 10^{+2} s. In GaAs, InP and various other compound semiconductors, recombination occurs with the emission of radiation, as discussed later in connected with light emitting diodes (LEDs).

3.10 The 'thick' diode: recombination and the diffusion length

Having neglected recombination in treating the 'thin' diode, we now include it, and find out how the concentration of injected carriers varies as they

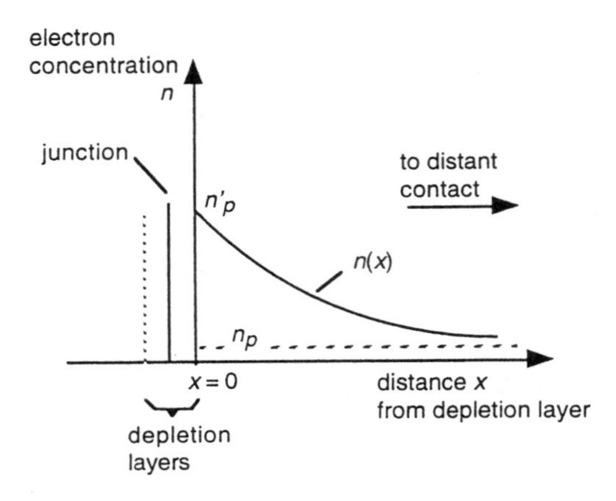

Fig. 3.8 Carrier concentrations in a 'thick' diode.

diffuse away from the junction in a 'thick' diode. Figure 3.8 shows how the concentration $n(x)$ of injected electrons is expected to fall with distance x from the edge of the depletion layer. We shall assume that the bias voltage across the junction maintains the constant concentration n_p' at $x = 0$, and that all of the bias voltage appears across the depletion region — i.e. the voltage drop across the neutral region remains negligible.

The shape of the distribution in Fig. 3.8 results from a dynamic equilibrium within any small volume element between diffusion of electrons into that volume, and recombination within it. Consider therefore this equilibrium in the thin slice of unit cross-sectional area and thickness δx shown in Fig. 3.9. The concentration in this slice stays constant, because the number of electrons diffusing into it from the left exceeds that leaving towards the right, by just the amount needed to balance recombination.

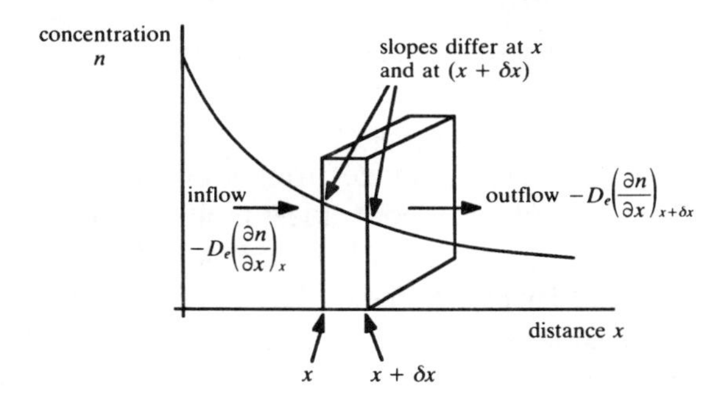

Fig. 3.9 Illustrating the calculation of recombination and diffusion in a 'thick' diode.

The rate of loss of electrons occuring within this volume by recombination is found using eqn. (3.23), and is

$$\frac{\partial n}{\partial t}\delta x = \frac{-(n(x) - n_p)\delta x}{\tau_e}$$

The inflow of electrons by diffusion across the plane x is $-D_e \partial n/\partial x$, while at the plane $x + \delta x$ the outflow is $-D_e[\partial n/\partial x + \delta x \partial^2 n/\partial x^2]$. The difference between them is a net influx of $D_e(\partial^2 n/\partial x^2)\delta x$ by diffusion. Equating the net influx to the recombination rate gives the equation

$$D_e \frac{d^2n(x)}{dx^2} - \frac{(n(x)-n_p)}{\tau_e} = 0 \tag{3.24}$$

This differential equation has the solution

$$n(x)-n_p = (n'_p - n_p)\exp\left[\frac{-x}{(D_e\tau_e)^{1/2}}\right] \tag{3.25}$$

as can be shown by substitution into eqn. (3.24).

PANEL 3.3

Direct and indirect band gaps

Recombination of electrons and holes in silicon or germanium almost invariably occurs by first trapping the hole or the electron at an impurity. This stems from a peculiar property possessed by the energy bands in these materials, which causes free electrons with energy near E_c to have much greater momentum (in silicon) or less momentum (in germanium) than holes with energy near E_v. (This can only be explained using quantum mechanics.) Then so-called RADIATIVE RECOMBINATION, involving the emission of a photon of energy nearly E_g, almost never occurs since momentum cannot be conserved during the process. For, if a free electron is to 'fill' a hole in Si, it must first lose its momentum. Although the photon can carry away a little momentum (an amount h/λ for a wavelength λ) it cannot remove enough for recombination to occur in this way. Instead, recombination occurs in two stages. First, the electron (for example) becomes trapped at an impurity atom, losing its momentum as it does so. Subsequently a hole arrives and the electron 'fills' it. Recombination accompanied by a change in momentum is called INDIRECT RECOMBINATION, and dominates in materials with energy bands similar to those in silicon or germanium. They are said to have an INDIRECT BAND GAP, to distinguish them from materials such as GaAs or InP, in which radiative recombination occurs easily. The latter are said to have a DIRECT BAND GAP.

This equation describes the shape of the graph in Fig 3.8.

Note that eqn. (3.25) shows that the excess concentration $(n(x) - n_p)$ falls by $1/e$ in a distance equal to $(D_e\tau_e)^{1/2}$. This distance is called the DIFFUSION LENGTH, for which we shall use the symbol L_e, of the electrons in the *p*-type material.

The diffusion current of electrons through the neutral *p*-type region of the diode is now dependent on x, falling as the slope of the graph $n(x)$ decreases with increasing x. For $x > 0$, the total current flowing must be partly carried by holes, since there must be continuity of current throughout the diode. The hole current must rise as x increases, until when $x \gg L_e$, it constitutes practically the entire current.

Where $x < 0$, no recombination of electrons can occur, so that the electron current at $x = 0$ represents the total current arising from the injection of electrons, and has the value

$$J_n = D_e e\frac{dn}{dx} = D_e e\frac{(n'_p - n_p)}{(D_e\tau_e)^{1/2}} \tag{3.26}$$

The final form here was obtained by differentiating eqn. (3.25) and putting $x = 0$.

The consequence of this equation is that the current of electrons is proportional to $(n'_p - n_p)$, just as in the 'thin' diode. The only difference is that the diffusion length $(D_e\tau_e)^{1/2}$ appears in the denominator, where previously it was the distance between junction and contact (compare eqn. (3.10)). The current no longer depends on the diode length, but on the diffusion constant and the minority carrier lifetime.

We can now appreciate the condition required to ensure that negligible recombination occurs in the 'thin' diode: the distance of the contact from the junction should be much less than the diffusion length.

Since the minority carrier lifetime is very variable, so diffusion lengths vary widely, even in good quality wafers.

3.11 Light emitting diodes (LEDs)

In a thick diode which is made of e.g. GaAs rather than of silicon, recombination occurs in the neutral region with emission of infra-red 'light'. The wavelength of the emitted radiation is obtained by equating the energy

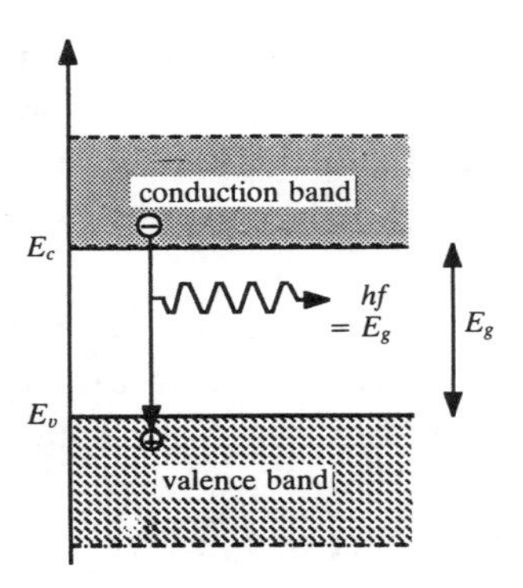

Fig. 3.10 Illustrating the process of recombination with radiation in a GaAs diode.

lost by the electron with the energy, hf, of the photon emitted. The process is illustrated schematically of the energy level diagram in Fig. 3.10.

Since the most probable energy for an electron is $kT/2$ above the band edge E_c, while that of a hole is similarly $kT/2$ below E_v, the wavelength, λ, of the most intense radiation is expected to be given by

$$hf = \frac{hc}{\lambda} = E_g + kT$$

Using this equation it is easy to show that a GaAs LED ($E_g = 1.4$ eV) should emit photons of wavelength about 0.86 µm — in the near infra-red. In practice, several minor effects contribute to an increase in the wavelength to about 0.9 µm. By replacing some of the gallium with aluminium, the energy gap can be increased enough to bring the wavelength into the visible spectrum.

The rate of emission of photons — and hence the intensity of the emitted light — is proportional to the product of p and n in the region close to the junction where recombination occurs (see section 1.10). To obtain a useful intensity of emission, the injected minority carrier density must be made very high. To achieve this, the doping is made very heavy, and the current density must be as large as heating effects will allow.

High current density makes the forward voltage across the device higher than otherwise, for two reasons. One is that the forward bias on the junction itself becomes almost equal to the built-in potential difference V_0. The other is that the series resistance R_s in the neutral regions of the diode causes an additional voltage drop IR_s. The sum of the two voltages is typically more than 1.5 V, and the current–voltage characteristic is then given by eqn. (3.13).

Another consequence of heavy doping is that the reverse breakdown voltage is very small — about 1 V — because the depletion layer is very thin. So LEDs must be protected against reverse voltages above about 1 V.

Most of the electrons injected across the junction recombine within the recombination length $L = (D\tau)^{1/2}$, typically 5 μm, so that nearly all the radiation is emitted close to the junction. The small recombination length is due to a short lifetime, τ. The lifetime, τ_e of electrons is small because the heavy doping has the following two effects: (i) free holes are frequently encountered by electrons, increasing the probability of recombination, (ii) holes trapped at dopant atoms also often recombine with electrons, but without emission of radiation. The latter process becomes more probable, the higher is the concentration of acceptor atoms available to trap holes. Thus the probability that recombination occurs *without* emission of radiation rises with doping, and sets a practical upper limit both to the amount of radiation emitted and to the useful doping level. So there is a practical upper limit to the doping levels. However, since the speed with which the radiation can be turned on or off depends on τ_e, an LED for high speed optical fibre communications may be so heavily doped that its overall efficiency is often quite low, perhaps 1 per cent.

Silicon and germanium cannot be used for making LEDs, because recombination in them occurs predominantly by the indirect process of first trapping, then recombining without emission of radiation. The probability of so-called radiative recombination is very low, for reasons discussed in Panel 3.3.

3.12 Photodiodes

All semiconductor devices which detect light rely on the fact referred to in Chapter 2, that an electron–hole pair is created by the absorption of a photon, provided that the photon has an energy at least as great as the energy gap. This allows an increase in any current flowing in the semiconductor, if the electron–hole pair do not recombine. To prevent this, an electric field is used to separate the electron and hole, by causing them to move off in opposite directions. Since a large field exists in a *p–n* junction even with no bias voltage, a p–n diode is the commonest device for achieving pair separation. The process is illustrated schematically in Fig. 3.12(a).

The current–voltage characteristic of such a diode under various levels of illumination is shown in Fig. 3.11. The additional current resulting from illumination is almost completely independent of bias, so the I–V curve is shifted down the current axis by a constant amount directly proportional to the intensity of the light. This proportionality is accurately maintained over an extremely wide range of intensities.

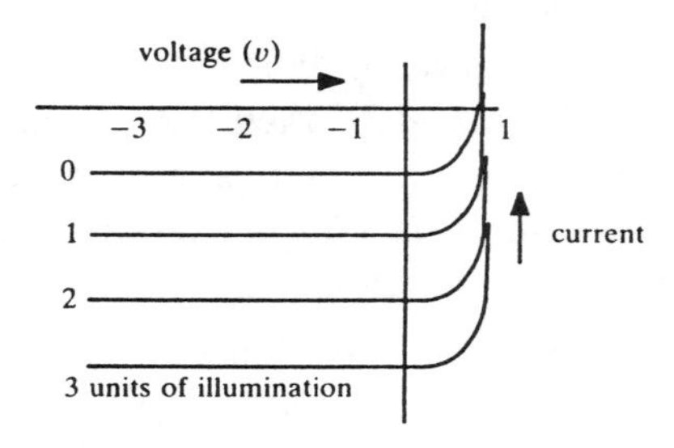

Fig. 3.11 Current–voltage characteristics of a photodiode under various levels of illumination.

The magnitude of the current flowing in reverse bias is governed by the proportion of electron–hole pairs which escape recombination before reaching the contact. If this proportion is η, the charge which flows in the external circuit is ηe for each photon collected. The power flow in a light beam of frequency f delivering N photons/s is Nhf watts, while the corresponding current generated in the diode is ηNe. The diode sensitivity can be expressed as the number of amperes of current per watt of light and is thus

$$\frac{\eta Ne}{Nhf} = \frac{\eta e}{hf}$$

To understand what controls the efficiency η of converting a flux of photons into electrons, consider Fig. 3.12(a). The efficiency with which electrons and holes reach the external circuit depends on whether or not the photon is absorbed at a point where the field is high, i.e. within the depletion layer. To maximise the chance of this, the depletion layer is made wider than normal by lowering the doping in the region of the junction itself. This region is therefore nearly intrinsic silicon, and the structure is referred to as a *p–i–n* diode (or PIN diode).

Note that the depth in Fig. 3.12(a) at which the photons are absorbed depends upon the wavelength of the light, as can be seen by studying Fig. 2.7. That figure shows how the average penetration depth of light varies with wavelength. Photons of longer wavelength, having energy equal to the energy gap E_g are weakly absorbed and penetrate to depths beyond the junction. The current they produce is small because the electron–hole pairs are not separated before recombination occurs, so the diode is not very

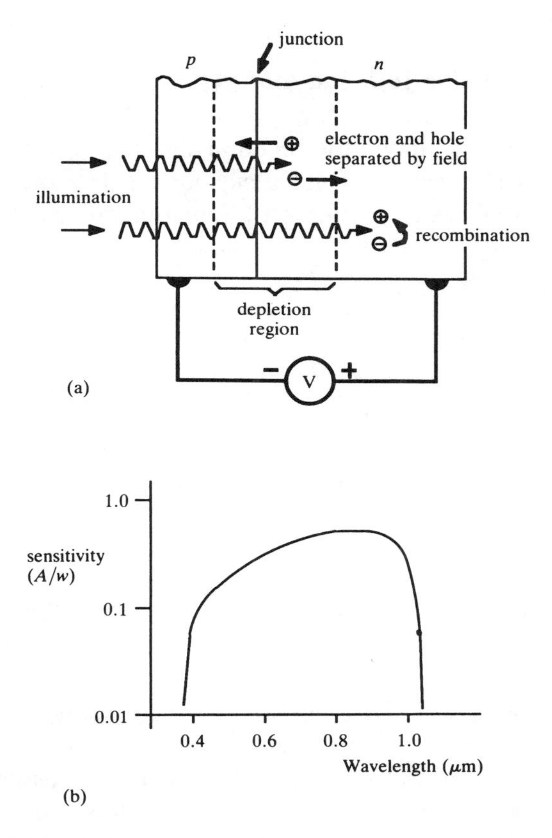

Fig. 3.12 (a) Creation of electron–hole pairs in a photodiode, shown in section. (b) Typical sensitivity in A/W of a silicon photodiode, plotted against wavelength.

sensitive at long wavelengths. At the other extreme, photons with energy well above E_g are strongly absorbed before they can penetrate to the depth of the depletion region. They, too, produce little current, as the electron–hole pairs do not separate quickly enough in the small field at the front surface. Again, the diode efficiency is low at these short wavelengths. At intermediate wavelengths, where the depth of penetration matches the junction, the diode acts as intended, and the sensitivity, when plotted against wavelength as in Fig. 3.12(b), displays a broad maximum. In practice the efficiency η of converting photons to electrons is about 50–80 per cent — both the premature recombination of electrons, and the reflection of part of the incident light at the surface account for the difference.

3.13 Circuit models for junction diodes

Two ways of modelling a diode for circuit analysis are given in this section: a 'small-signal' model suitable for hand calculations, and a more comprehensive, large-signal model as used in computer simulations of circuit behaviour.

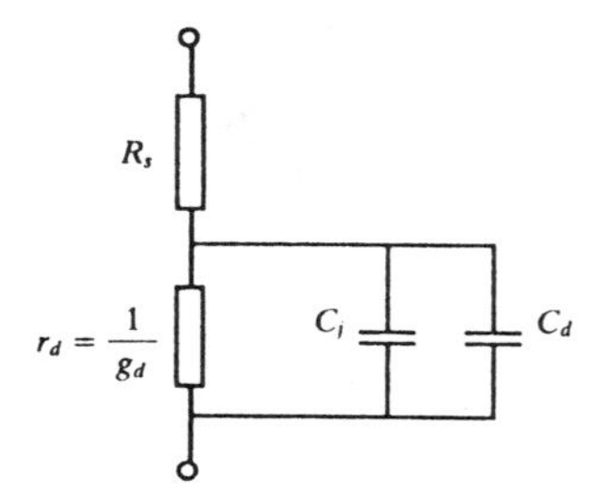

Fig. 3.13 Small-signal equivalent circuit of a diode.

When the diode is used in a circuit in which small alternating signal voltages and currents are superimposed on static (d.c.) values, the diode can be replaced for the purposes of calculating signal voltages and currents by the SMALL-SIGNAL EQUIVALENT CIRCUIT shown in Fig. 3.13.

Let the static voltage across the junction be V, and treat the a.c. signal as an infinitesimal voltage dV. The total current is similarly expressed as the sum of I and dI. The equivalent a.c. conductance g_d is the ratio dI/dV, which is obtained by differentiating the I–V relation given by eqn. (3.11):

$$g_d = \frac{dI}{dV} = \frac{e}{kT} I_s \exp \frac{eV}{kT} \tag{3.27}$$

The inverse of g_d is the resistance of the equivalent resistor r_d in Fig. 3.13, whose value according to eqn. (3.27) depends on the d.c. component V of the diode voltage. When V is more than a few hundred millivolts, the factor $I_s \exp(eV/kT)$ very nearly equals the d.c. diode current I, so that

$$g_d \cong eI/kT \text{ (FORWARD BIAS)} \tag{3.28}$$

The series resistance R_s discussed in (b) of section 3.6 is also included in Fig. 3.13. It is often negligibly small compared to r_d.

In reverse bias, leakage effects become important in determining the equivalent resistance, and the equivalent capacitance has a much lower impedance than the resistor.

The two capacitors in Fig. 3.13 represent the two capacitive effects referred to in section 3.8. C_d is a current-dependent capacitance which models the storage of injected carriers in the neutral, or diffusion, region of the diode. It is called the DIFFUSION capacitance, or STORAGE capacitance, and is discussed in section 3.14. C_j is the depletion-layer, or junction capacitance given by either eqn. (3.19) or (3.20). Both capacitances depend on the applied voltage or current in such a way that in forward bias $C_d \gg C_j$, while in reverse bias $C_j \gg C_d$ as seen in Fig. 3.14.

An alternative way of modelling a diode is to use the full I–V relation of eqn. (3.12). This approach is used in the standard computer simulations, of which the best known is probably **SPICE** (**S**imulation **P**rogram **I**ntegrated **C**ircuit **E**mphasis), which is widely used for simulating the behaviour of integrated circuits.

To specify the diode the values of the following parameters must be fed into the program.

I_s – the saturation current of the diode at 25°C

η – the correction factor for the exponent in eqn. (3.12)

V_0 – the built-in voltage given by eqn. (3.2), which is determined by the doping levels

C_{j0} – the depletion layer capacitance at zero bias, given by eqn. (3.20) when $V = 0$

m – the exponent in the capacitance eqn. (3.20), which depends on the doping gradient in the depletion region. With C_{j0}, above, this enables the depletion layer capacitance to be calculated at any bias voltage

R_s – the series resistance of the neutral regions and of any contacts to the external circuit. It may often be assumed zero

τ_t – the so-called transit time which is proportional to the diffusion or storage capacitance C_d shown in Fig. 3.13. See eqn. (3.33).

3.14 Diffusion or storage capacitance

In forward bias, the excess charge stored in the neutral, or DIFFUSION REGION, of the diode leads to a delay whenever an attempt is made to change the voltage across the junction. Because this region is electrically neutral, there is not only an excess of minority carriers (already shown in Fig. 3.3) but also an equal excess of majority carriers, so that their charges balance and the region is neutral. All of these carriers must be re-adjusted in number when, for example, the external circuit causes an alternating voltage to appear across the diode. This results in a flow of charge in and out of the diode which is modelled by the DIFFUSION CAPACITANCE sometimes called the STORAGE capacitance. This small-signal capacitance can be calculated from the change δQ of the excess minority carrier charge Q, stored in the neutral diffusion regions of the diode, which accompanies a small change δV in the applied voltage. We shall concentrate on studying the case of the *thin* diode.

The excess of electrons on the p-side of the junction is related to the hatched area in Fig 3.3: i.e. to the area between the graph of the concentration $n(x)$ of electrons in the diffusion region and the line $n = n_p$. If the applied voltage increases to $V + \delta V$, this area increases, since n'_p rises. The resulting change δQ_e in the stored electron charge can be expressed as $\delta Q_e = (dQ_e/dV)\delta V$.

Q_e can be calculated with the aid of Fig 3.3: it is the product of the volume of the diffusion region with the mean excess charge concentration, which is just half the maximum excess concentration $(n'_p - n_p)e$. For a diode of cross-sectional area A:

$$Q_e = \frac{1}{2}(n'_p - n_p)eL_n A = \frac{1}{2}L_p eAn_p[\exp(eV/kT) - 1] \qquad (3.29)$$

A similar expression can be derived for the charge stored by holes in the diffusion region on the n-type side of the junction.

Now the forward current due to electrons is also proportional to the expression $n_p[\exp(eV/kT) - 1]$, as can be seen in eqn. (3.10). This equation can be used then to express Q_e in terms of the electron current I_e:

$$Q_e = \left(\frac{L_p^2}{2D_e}\right) I_e \qquad (3.30)$$

Adding the corresponding expression for the charge due to holes stored on the other side of the junction to that for Q_e gives the total stored charge Q_{st} in terms of the total current I:

$$Q_{st} = \left\{\left(\frac{L_p^2}{2D_e}\right) I_e + \left(\frac{L_n^2}{2D_h}\right) I_h\right\} = \left\{\left(\frac{L_p^2}{2D_e}\right) f + \left(\frac{L_n^2}{2D_h}\right)(1-f)\right\} I$$

where I_h is the current carried by holes, and f is the fraction of the total current which is carried by electrons, so that $f = I_e/(I_e + I_h)$.

Now we shall differentiate this equation with respect to voltage, to find the diffusion capacitance C_d

$$C_d = \frac{dQ_{st}}{dV} = \left\{\left(\frac{L_p^2}{2D_e}\right) f + \left(\frac{L_n^2}{2D_h}\right)(1-f)\right\} \frac{dI}{dV} \qquad (3.31)$$

But dI/dV is just the diode conductance g_d given in eqn. (3.27), so that we can write

$$C_d = \left\{\left(\frac{L_p^2}{2D_e}\right) f + \left(\frac{L_n^2}{2D_h}\right)(1-f)\right\} g_d \qquad (3.32)$$

So the small-signal diffusion capacitance is directly proportional to the current through the junction. It is also dependent on the quantities $L_p^2/2D_e$ and $L_n^2/2D_h$, which have the dimensions of time and are known as TRANSIT TIMES. They are the average times taken respectively by an electron or hole to travel the length L_p or L_n of the diffusion region (see section 3.15 below).

By defining an *effective* transit time τ_t as the weighted sum of these two times, eqn. (3.32) can be expressed very simply:

$$C_d = \tau_t g_d \tag{3.33}$$

Compare this equation with eqn. (3.32) to see that τ_t is equal to the expression inside the { } brackets in the latter equation.

Removal of the stored charge Q_{st} causes a noticeable delay when the diode is switched off. This effect is discussed later in section 3.16.

Now let us compare the magnitudes of the diffusion and junction capacitances. The first thing to note is that both rise as the bias voltage increases in the forward direction. However, C_d increases the more rapidly, since in eqn. (3.33) g_d rises exponentially with voltage, whereas eqn. (3.19) shows C_j rising as $(1 - V/V_o)^{-1/2}$. Figure 3.14 illustrates the voltage dependence, showing that with a *forward bias* of more than a few hundred millivolts the diffusion capacitance normally dominates, and C_j is often negligible. The converse applies for reverse bias voltages.

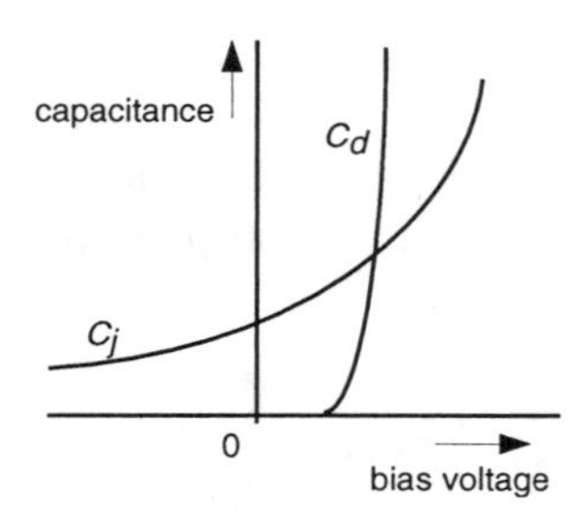

Fig 3.14 Voltage dependence of the small-signal junction capacitance C_j and the small-signal diffusion capacitance C_d of a junction diode.

3.15 The transit time

We shall now explain why the quantities $L_p^2/2D_e$ and $L_n^2/2D_h$, defined above, are called transit times.

Since any individual carrier in the diffusion process moves entirely randomly, its *average* velocity is exactly zero, and the time it takes to traverse the distance L_n at zero velocity is infinite! However the *mean square* distance $\overline{x^2}$ travelled in time t is non-zero, and can be shown using statistical methods to be equal to $2Dt$. If we *define* the transit time for diffusion using this result, we find that it equals the ratio of stored charge to current flowing, exactly as is the case with a drift current. Thus from equation (3.30) above:

$$\tau_t = \frac{L_n^2}{2D_e} = \frac{Q_e}{I_e} \tag{3.34}$$

Typical values are given here:

with $L_n \sim 20\ \mu m$ (e.g. a discrete device), $D_h = 0.025\mu_h = 10^{-3}\ m^2/s$: $\tau_t = 0.2\ \mu s$

with $L_p \sim 2\ \mu m$ (integrated circuit diode), $D_e = 0.025\mu_e = 3 \times 10^{-3}\ m^2/s$: $\tau_t = 1.3$ ns

The transit time τ_t is a *fundamental parameter* limiting the speed of switching of a thin diode. Indeed, the concept of a transit time will crop up later in connection with transistors, and is also relevant to the performance of vacuum devices such as the cathode ray tube, and any switch which relies on *flow* for its operation.

3.16 Large-signal switching

Now consider what happens when we try to switch a thin diode rapidly between *on* and *off* states, using the circuit and input voltage waveform shown in Fig 3.15, where the voltages V_1 and V_2 are each several volts in magnitude. The turn-on and turn-off transitions must be considered separately.

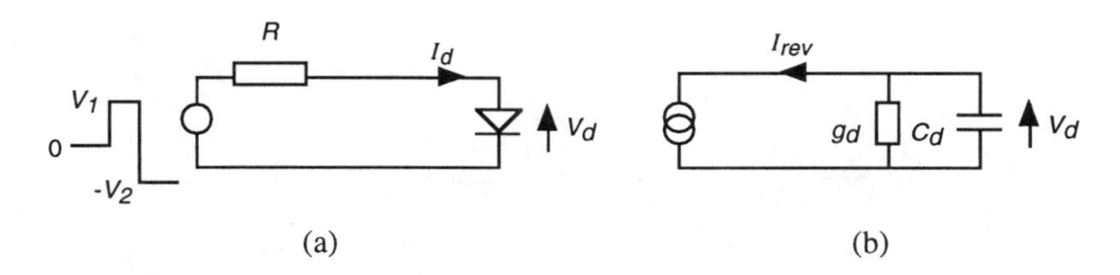

Fig 3.15 (a) The circuit used to discuss the speed of switching of a junction diode. (b) Equivalent circuit for analysis of turn-off transient.

Turn-on

The voltage source and series resistance in the circuit in Fig 3.15(a) act together approximately as a constant current source, if $V_d \ll V_1$. Although

V_d does not rise instantly to 0.6 V, because a finite time is needed to inject minority carriers, the current source forces instant current turn-on, as seen in the lower waveform plotted in Fig. 3.16(b), below the input voltage waveform. The rise of V_d to 0.6 V subsequently causes a small drop in current following the rising step, to a final value $(V_1 - V_d)/R = (V_1 - 0.6)/R$.

Since the current has changed, a different quantity of charge will ultimately be stored in the diode. The new value of stored charge must take some time to be established (otherwise an infinite current would flow), and this time delay is one of the important factors limiting the speed of circuits in which diodes are used.

The shape of the minority carrier distribution on each side of the junction will develop with time as illustrated in Fig. 3.16(c), beginning with the horizontal line labelled $t < 0$, and ending with the line marked $t \rightarrow \infty$. Note that, as the graph of p_n' rises, the slope dp/dx at the depletion layer edge is directly proportional to the current flowing, and hence is practically the same at all times during the transient.

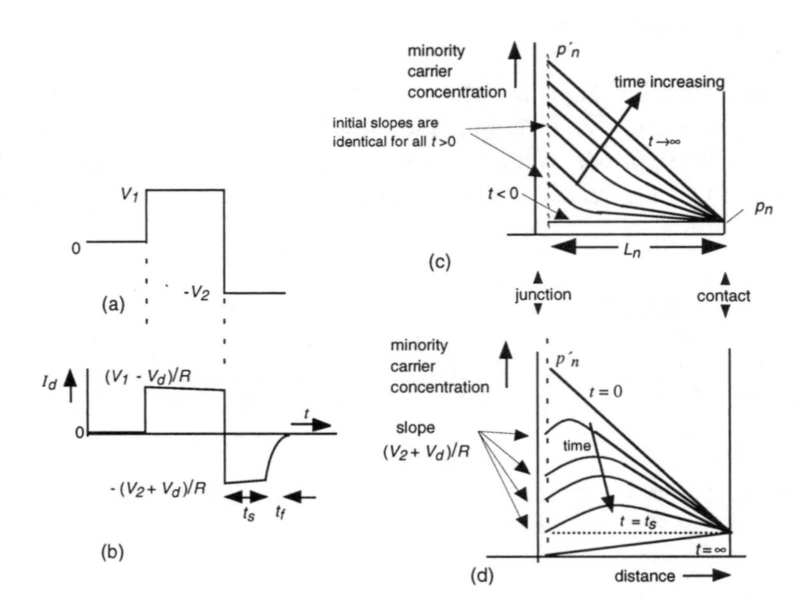

Fig. 3.16 In a diode, the input voltage waveform shown at (a) produces the current waveform at (b) below it. (c) shows the distribution of minority holes at various times during the turn-on transient (d) shows the collapse of the minority carrier concentration in the neutral region during turn-off.

Turn-off

At the beginning of turn-off, the distribution of minority carriers across the neutral region of the diode is shown in Fig 3.16(d) by the line marked $t = 0$. (We shall restart the clock for the discussion of turn-off). The voltage V_d across the diode is directly related to the carrier density p_n' (see eqn. (3.8)), and can fall only if p_n' itself falls. According to eqn. (3.8), V_d falls in proportion to $\log_e p_n'$. Hence V_d remains close to 0.6 V until p_n' has reduced substantially. This implies that an almost constant current flows in the circuit, of value $I_{rev} = -(|V_2| + 0.6)/R$ in the *reverse* direction through the diode! This current continues to flow until p_n' has fallen at least enough to equal p_n, the value in the unbiased diode, at which point $V_d = 0$ (see eqn. (3.8)), and the current has merely fallen in magnitude to $-|V_2|/R$.

The time during which the reverse current remains nearly constant at the above value is called the *storage time* t_s. The minority carrier distribution at the end of this time is sketched as the line $t = t_s$ in Fig. 3.16(d). The storage time is followed by a fall of the magnitude of the current back to zero with a fall time t_f, during which the remaining amount of charge either leaves or recombines within the diode*. These times are shown on the complete current waveform sketched in Fig. 3.16(b).

In order to estimate the time taken for all of the excess charge to be removed from the diode, we shall assume that the diode behaves similarly, whether the current changes are small or large. Now, if I_{fwd} and I_{rev} are small, we may treat the diode as if it has the small-signal equivalent circuit derived earlier. In the case that I_{fwd} and I_{rev} are not small, we assume that it is possible to use average values of the parameters needed to represent the diode properties, in order to get a reasonably close estimate of the true switching time. Secondly, we shall for simplicity assume that the resistor and voltage source in Fig. 3.15(a) can be replaced by an ideal current source, as shown in Fig. 3.15(b). The use of a current source means that the 'tail' of the current waveform in Fig. 3.16(b), during which the current falls because the source has a finite impedance, is no longer a separate part of the waveform, since the current is forced to remain constant throughout. The

---

*The fall time t_f is normally taken to end when the external current reaches 10% of I_{rev}, rather than when the charge reaches zero. We shall ignore the difference here, in order to simplify the calculations below.

switching time we calculate is thereby reduced somewhat below the true value, because the fall time t_f will be underestimated.

Whether the current I_{fwd} is small or large, the initial value $Q_{st}(0)$ of the stored charge may certainly be calculated in terms of the current, by use of eqn. (3.34). Thus, at time $t=0$, the charge is $Q_{st}(0)=I_{fwd}\tau_t$. On reversal of the current, the capacitor C_d discharges, partly through the conductance g_d and partly through the external current source. The rate of discharge of C_d is equal to the sum of the two currents flowing in those components, as follows:

$$C_d\frac{dV_d}{dt}=\frac{dQ_{st}}{dt}=-g_dV_d-I_{rev}$$

Since we know that $Q_{st}=C_dV_d$, and $C_d=g_d\tau_t$ this equation can be rearranged to give:

$$\frac{dQ_{st}}{dt}=-\frac{Q_{st}}{\tau_t}-I_{rev} \tag{3.35}$$

Note that the time constant τ_t which appears here can be seen from its definition in eqn. (3.34) to be dependent only upon the diode's dimensions and the diffusion coefficient of the minority carriers, in spite of the fact that both C_d and g_d depend upon the current flowing. It seems that τ_t may be treated as constant. However, equation (3.34) was derived assuming the minority carrier distribution had a simple triangular shape, which Fig. 3.16(d) shows is not true during the switching operation. Nevertheless, to take account of variations in τ_t would complicate the problem considerably. Assuming, for simplicity, that τ_t is constant, the solution of the differential equation (3.35), subject to the initial condition that $Q_{st}(0)=I_{fwd}\tau_t$, is

$$Q_{st}(t)=\tau_t(I_{fwd}+I_{rev})\exp(-t/\tau_t)-\tau_tI_{rev} \tag{3.36}$$

Thus the charge $Q_{st}(t)$, which resides in the capacitor C_d in Fig. 3.15(b), is expected to decrease exponentially with time, with the time constant τ_t. Note that eqn. (3.36) predicts that it decays asymptotically towards a new equilibrium value, equal to $-\tau_tI_{rev}$. This charge value can never be reached, unless we choose the magnitude of I_{rev} to be small enough that it is less than the reverse saturation current I_s. If τ_tI_{rev} happens to be larger than I_s, the exponential decay simply ceases as soon as the reverse current reaches I_s.

The stored charge thus reaches its final value (approximately zero) at this time. The time-dependence of the charge predicted by eqn. (3.36) is illustrated in Fig. 3.18, in which it is assumed that I_{rev} is large in magnitude compared to I_s.

We can see from Fig. 3.16(b) that the storage time t_s ends *before* the charge reaches zero, because further charge flows out during the fall time t_f. Assuming that all of the charge $Q_{st}(0)$ disappears in a total switching time t_{sw}, we can estimate this time by setting $Q_{st}(t_{sw}) = 0$ in equation (3.36), leading to the result:

$$t_{sw} = \tau_t \ln\left(1 + \frac{I_{fwd}}{I_{rev}}\right) \tag{3.37}$$

To recapitulate, we assume that this result gives an estimate of the total recovery time $t_s + t_f$ of Fig. 3.16(b), and is a good approximation, provided only that the value of τ_t which we use is a suitably averaged one. Although the simple model of the diode we have used predicts no change in τ_t with current (see eqn. (3.34)) we cannot expect this result to be particularly accurate. A more sophisticated analysis of this problem confirms that equation (3.37) gives a value for t_{sw} which is within about 50% of the correct value, except when I_{rev}/I_{fwd} becomes very large (greater than about 50), when the true recovery time becomes substantially smaller than eqn. (3.37) predicts. In many practical situations, (3.37) suffices as a first estimate.

A much more thorough analysis of the approximations involved is to be found in the book by Tyagi cited in the list of further reading.

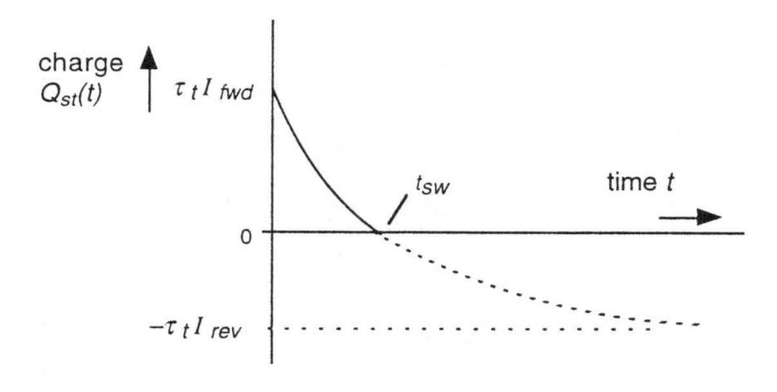

Fig. 3.17 The decay of charge in the storage capacitance at a function of time, according to equation (3.36).

3.17 Capacitances and switching times in the thick diode

The differences between the thick diode and the thin are discussed in this section.

The *junction* capacitances of the thin and thick diodes are identical when expressed in terms of the widths of the depletion layers, since the value does not depend on the total length L_n or L_p of the n-type and p-type regions.

The *Storage* or *Diffusion* capacitance is calculated from the minority carrier concentration graph in a similar manner to that used for the thin diode. The exponential shape of this graph, repeated in Fig. 3.18 for the case of a p^+n diode, was explained in section 3.10. The area under the graph of $p_n(x)$ is proportional to the stored charge Q_h due to holes. The charge Q_e stored on the p-type side of the junction is much smaller, owing to the few electrons available on the n-type side for injection across the junction. We assume that Q_e is small enough to be negligible.

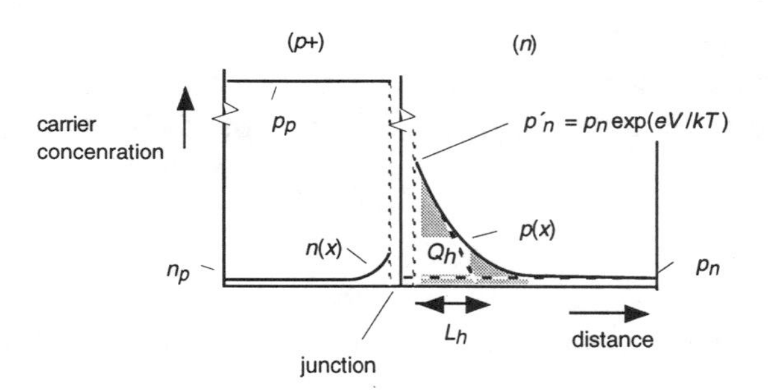

Fig. 3.18 Showing the distribution of stored minority carriers in the neutral regions of a thick p^+-n diode.

Following the method used for the thin diode, we can write

$$C_d = \frac{dQ_{\text{st}}}{dV} \cong \frac{dQ_h}{dI} \cdot \frac{dI}{dV} = \frac{dQ_h}{dI} \cdot g_d$$

But in this case it can be shown using Fig. 3.18 and eqn. (3.23a) (see Panel 3.4), that

$$\frac{dQ_h}{dI} = \frac{L_h^2}{D_h} = \tau_h \tag{3.38}$$

PANEL 3.4

All of the holes flowing into the neutral n-type region of a p-n diode, by injection across the junction, recombine there with electrons. Everywhere in this region, the rate of loss of holes by recombination in any infinitesimal volume δV must equal the rate of inflow of holes into that volume. Now by using eqn. (3.23a) for the rate of recombination per unit volume, we may compute the rate of loss of charge carried by the holes in a volume δV as follows:

$$e\frac{dp}{dt}\delta V = -e\left(\frac{p - p_n}{\tau_h}\right)\delta V = \frac{\delta Q_h}{\tau_h}$$

where δQ_h, defined in this equation, is the charge carried by the excess holes in the volume δV.

Now the rate of loss of *charge* by recombination expressed in the above equation must equal the rate of inflow of charge into the volume δV, i.e. the current δI into it. Hence $\delta I = \delta Q_h/\tau_h$

Since this is true for any element of volume, it must be true for the whole volume of the neutral n-type region, into which the total hole current I_h flows and recombines, i.e.

$$I_h = \frac{Q_h}{\tau_h}$$

where Q_h, as before, is the total charge stored by holes in the neutral n-type region. Equation (3.38) follows immediately from this result.

Since exactly the same argument can be applied to electrons entering the neutral p-region, the total diode current can be expressed in terms of an *effective lifetime* τ_L, defined exactly as was the effective transit time in eqn. (3.32), i.e. by the equation

$$I_h = \frac{Q_{st}}{\tau_L}$$

where $\tau_L = [\tau_e f + \tau_h(1 - f)]$, and f, as earlier, is the fraction of current carried by electrons.

where τ_h is the hole *lifetime*, in place of the *transit time* in the case of the thin diode*. Thus

$$C_d = \tau_h g_d \quad \text{(thick } p^+n \text{ diode)} \tag{3.39}$$

So the expression for diode diffusion capacitance given in eqn. (3.33) still applies, *as long as the transit time is replaced by the appropriate lifetime* (i.e. the electron lifetime in an n^+p diode, the hole lifetime in a p^+n diode).

Since large-signal switching is controlled by the diffusion capacitance, it occurs in a similar way to switching of the thin diode, although the shape of the minority carrier distribution during switching differs in detail. The storage time, during which the diode remains forward biased, and the total recovery time are controlled by the *lifetime*, which replaces the *transit time* in equations (3.35)–(3.37).

The similarity to the thin diode enables a computer package such as SPICE to model either kind of diode with a single set of equations: the diode length and the time constants must merely be chosen to suit by the user.

3.18 Schottky diodes and ohmic contacts

Types of metal-semiconductor contact

Any two dissimilar materials placed in contact will exchange mobile charge. Any metal thus falls into one of two categories relative to any particular piece of doped (or undoped) semiconductor. It either (a) donates electrons to the semiconductor, or (b) accepts electrons from the semiconductor, until the Fermi levels become equal. The class into which a metal falls may differ according to the doping concentration in the semiconductor and, more importantly the doping type.

Placing a particular metal in contact with an *n-type* semiconductor results in either an *accumulation region* — case (a) above — or a depletion region — case (b) — at the surface of the semiconductor, giving one or other of the two possible energy level diagrams shown in Fig. 3.19(a) or (b). Note that both of these are sketched for the situation in which no bias voltage is applied.

*Note particularly that the factor 2 in the definition of the transit time in eqn. (3.32) does not appear in the denominator of eqn. (3.38).

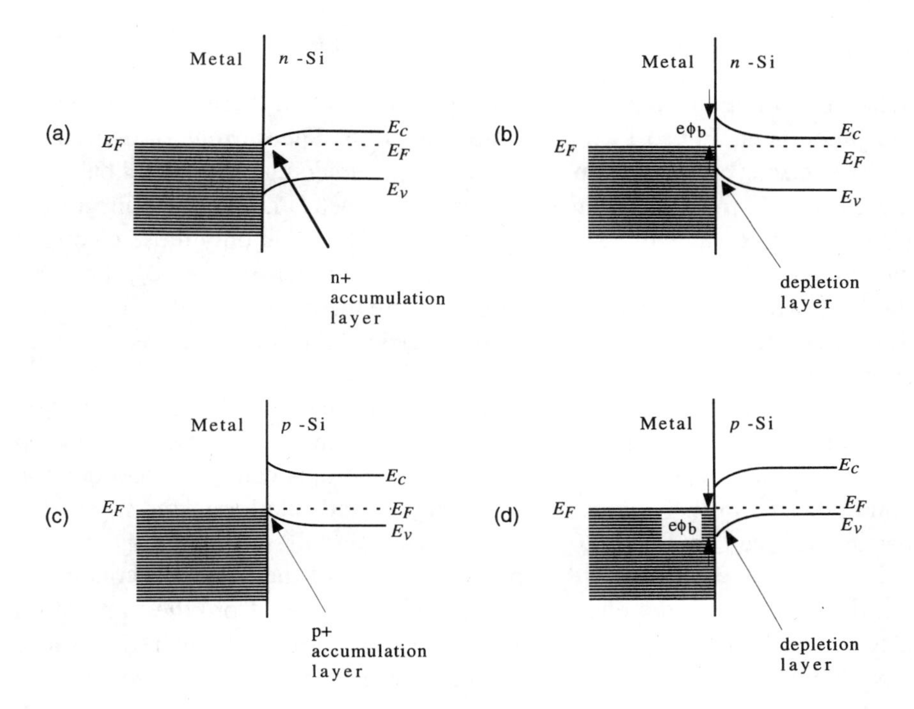

Fig. 3.19 The Energy level diagrams of (a) a contact in which a metal by its nature charges *positive* relative to an n-type semiconductor, and (b) a contact in which the metal charges *negative*. (c) and (d) show the same for a p-type semiconductor.

Most useful metals when placed in contact with n-type silicon form a depletion layer, i.e. the energy level diagram is as shown in Fig. 3.19(b). The current-voltage relationship in such a contact is a rectifying one: it is usually called a SCHOTTKY DIODE. The energy difference $e\phi_b$ is commonly between 0.5 eV and 1.0 eV. The alternative, Fig. 3.19(a), leads to ohmic, i.e. linear, electrical behaviour. Such a contact is termed OHMIC.

Metals in contact with p-type silicon similarly cause either an accumulation or depletion region for holes. Again, the common metals usually produce a rectifying contact, associated with depletion of holes near the interface. The corresponding energy level diagrams are those in Figs. 3.19(c) and 3.19(d).

We shall concentrate on the n-type situation here in explaining the current-voltage behaviour, and we consider first the Schottky diode.

Explanation of current flow — the Schottky diode rectifier

When no voltage is applied, free electrons in random thermal motion cross the junction in Fig. 3.19(b) in either direction. The current I_R carried by those electrons crossing from the metal to the semiconductor exactly balances the current I_F from the semiconductor to the metal, i.e. the net current $I = (I_F - I_R)$ is zero since there is no bias voltage. Note that only those electrons can cross the boundary which have an energy of at least $(E_F + e\phi_b)$, i.e. only those above the peak energy level reached by the conduction band edge E_c. The step $e\phi_b$ in energy thus acts as a barrier to those electrons of lower energy. It is termed a SCHOTTKY BARRIER.

When a bias voltage is applied, the extra potential difference appears across the depletion layer only, just as in a *p–n* junction, because the depleted region has the highest resistance. Therefore the bias causes the conduction band edge E_c in the semiconductor to rise or fall relative to the top of the barrier, as seen in Fig. 3.20(a) or (b). The bias thus changes the barrier height seen by electrons in the conduction band of the bulk semiconductor. On the other hand, the energy barrier height $e\phi_b$ seen from the view of an electron at the Fermi level in the metal remains virtually unaltered, because any change there would have to result from a potential difference between the two layers of atoms lying either side of the junction, about 0.3 nm apart. The total voltage applied, perhaps a few volts, must be equally shared between perhaps 300 interatomic spacings lying in the high-resistance, depleted region, resulting in no more than about a millivolt between each pair of atoms.

The two cases of positive and negative bias voltage are now considered in turn.

Consider first the case of a negative bias of magnitude $-V$ volts applied to the metal relative to the bulk semiconductor, illustrated in Fig. 3.20(a). The energy levels in the metal are all raised relative to those in the semiconductor by the amount eV.

Electrons attempting to flow from from the semiconductor to the metal see an increased barrier height $e(\phi_b + |V|)$, making flow more difficult in that direction. Remember that thermal energy is needed to lift free electrons above the conduction band edge at E_c. The current I_F thus decreases, in proportion to $\exp(-eV/kT)$, and is readily reduced by many orders of magnitude below its value at zero bias, with only a few hundred millivolts of bias voltage.

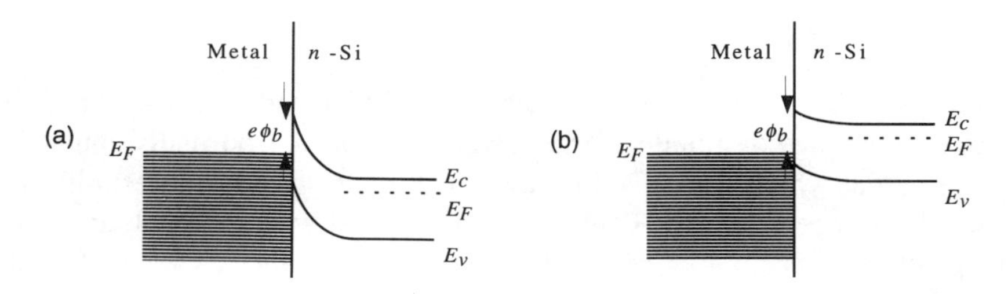

Fig. 3.20 Energy level diagrams of a Schottky diode under (a) reverse bias, and (b) forward bias.

Electrons flowing from from the metal to the semiconductor see an essentially unchanged barrier height $e\phi_b$, for the reason given above. So the current I_R is unchanged by the bias voltage, and the net current is $I_F - I_R \cong - I_R$. But I_R is a rather small current, since only those electrons with enough thermal energy (greater than $e\phi_b$) can cross the barrier. Their numbers depend on the factor $[\exp(-e\phi_b/kT)]$, which is typically 10^{-14} when ϕ_b equals 0.8 V. The net current flow $I_F - I_R \cong -I_R$ is thus very small, its magnitude dependent mostly on the barrier height $e\phi_b$, and not on the applied voltage. Hence I_R represents the reverse saturation current of the diode.

Now consider a positive bias $+V$ applied to the metal, relative to the semiconductor. Figure 3.2(b) shows the resulting energy levels. Note the significant reduction in barrier height $(\phi_b - V)$ seen by electrons flowing from the semiconductor into the metal. The number of carriers with sufficient thermal energy to cross the barrier into the metal is proportional to

$$\exp[-e(\phi_b - V)/kT] = \exp(-e\phi_b/kT)\exp(eV/kT) \propto \exp(eV/kT)$$

Hence the current $I_F \propto \exp(eV/kT)$ is substantially raised, and the net current through the diode becomes

$$I = I_F - I_R \propto [\exp(eV/kT) - 1]$$

The diode conducts readily, and we refer to this as the *forward* direction of current flow, as for a *p–n* junction diode. The forward voltage drop at a representative current density depends upon the barrier height $-e\phi_b$. As an example, a diode of aluminium in contact with *n*-type silicon has a typical forward voltage drop of about 0.25 V, or slightly less if the doping concentration in the silicon is low.

Capacitance of the Schottky diode

Electrons crossing to the metal under forward bias can flow away readily: there is no storage of mobile charge. Switch-off speed is normally controlled by the rate at which the depletion-layer width is restored to its equilibrium value. This depends on the capacitance associated with the depletion layer (its junction capacitance), and on any series resistance. This makes the Schottky diode a much faster switch than a junction diode, because the junction capacitance can be made small by widening the depletion layer. Ultimately another limit is reached, set by the speed at which electrons cross the depletion region. Since electrons cannot move faster than a velocity of about 10^5 m/s in silicon (the saturation velocity), the minimum transit time across a 1 μm thick depletion region is about 10 picoseconds. As the depletion layer widens, its capacitance falls, but the transit time rises, and eventually the latter limits the switching speed.

Because Schottky diodes can switch much faster than *p–n* diodes, the reader may wonder why they are not more widely used. The reason lies in part in the difficulty found in practice, of making a reliable contact of high perfection, between two materials of such dissimilar nature as a metal and a covalently bonded semiconductor. Schottky diodes therefore tend to be more expensive, and also have lower reverse breakdown voltages than *p–n* diodes.

Current flow in ohmic contacts

In the situation which is shown in Fig. 3.1(a), there is no barrier to the flow of electrons in either direction for bias of either sign, since there is no depleted, high resistance region at the junction. Indeed, the accumulation region contains a higher concentration of electrons than does the bulk semiconductor, as the Fermi level is nearer to E_c. Flow is limited only by resistance of the bulk semiconductor, for the metal usually has lower resistivity.

In the case of most metals used in practice for making contacts to semiconductors, this ideal situation is not easily achieved. However "ohmic" contacts are often made in another way. For example, aluminium which is in contact with very heavily doped silicon creates a Schottky barrier which is so thin (<10 nm) that electrons can "tunnel" through it. This happens because of the wave-like property of electrons discussed in Chapter 2. The contact behaves for practical purposes as an ohmic contact. Thus where aluminium metal is required to make an ohmic contact to silicon, the silicon is locally doped *n*+ or *p*+ by implanting additional dopants, to create a thin enough Schottky barrier.

Metals in contact with *p*-type semiconductors can be discussed in exactly the same way as above, holes taking the role of electrons within the semiconductor. Both rectifying and ohmic contacts may be made, and readers are invited to consider for themselves how the same principles as those used here can be applied to these cases.

Summary of New Terminology

The BUILT-IN VOLTAGE is the potential difference across a junction in zero bias conditions.

A DEPLETION LAYER is a region depleted of mobile charge carriers.

JUNCTION CAPACITANCE is associated with changes in width of the depletion regions.

DIFFUSION CAPACITANCE is due to storage of minority carriers in the neutral, diffusion regions.

The TRANSIT TIME of a diode is the quantity relating charge stored in the diffusion region and forward current. In a short, asymmetrically doped diode, it equals the average time for the dominant minority carrier to cross the diffusion region.

SPICE is one of the better-known suites of computer modelling programs which includes diode and transistor models.

A VARACTOR DIODE is a diode designed for use as a voltage-controlled capacitor.

The DIFFUSION LENGTH is the $1/e$ decay length of the concentration of injected minority carriers.

The LIFETIME of minority carriers is the average time of survival of excess minority carriers before recombination occurs.

PROBLEMS

3.1 Is a depletion layer depleted of (a) dopant atoms (b) electrons (c) majority carriers (d) minority carriers (e) charge?

3.2 Why it is that in eqn. (3.1) the flows of electrons and holes must be separately equated to zero, rather than merely the *total* current?

3.3 Sketch the charge distribution in an unbiased junction, and relate it to the built-in potential difference across the junction and the electric field distribution in it.

3.4 Justify the assumption that the net hole and electron flows may be assumed to be nearly zero even when the junction is forward biased.

3.5 Explain why the concentration of minority carriers on either side of a junction is raised by applying a forward bias. Why is the concentration of majority carriers barely altered?

3.6 Is the diffusion length:
(a) the distance between the junction and the depletion layer edge
(b) the distance between the contact and the depletion layer edge
(c) the average distance electrons diffuse from the depletion layer edge before recombining?

3.7 Is the lifetime of electrons in a *p*-type semiconductor
(a) the average time it takes them to diffuse to the contact
(b) the average time before they recombine
(c) the average time they take to cross the depletion layer?
What factors determine the lifetime?

3.8 A junction has a built-in potential difference of 0.8 V and a forward bias of 0.55 V is applied. What is the potential difference across the depletion regions? What is its value if the bias is reversed?

3.9 Compare the built-in voltages in junctions made in Si and Ge having the doping concentrations $N_D = 10^{20}$ and $N_A = 10^{23}$, at 300 K.

3.10 The junctions in Problem 3.9 are made in diodes with area 0.1 mm^2 and with ohmic contacts at distances of 0.1 mm on either side of the junction, Assuming negligible recombination, compare the reverse saturation currents in these diodes. List in order of importance the factors contributing to the difference between them.

3.11 Would the currents in Problem 3.10 be larger or smaller if recombination could not be neglected?
Why is the measured reverse current in a silicon diode very much larger than calculated?

3.12 The neglect of recombination in Problem 3.10 implies that the electron lifetime is long. Compared to what value must it be long?

3.13 Find the depletion layer widths in the diodes of question 3.9 at zero bias. The permittivity of Ge is $16\varepsilon_0$.

3.14 Calculate the forward bias on a junction needed to raise the injected minority carrier concentration by a factor of 10^5. Find the forward currents in the diodes in Problem 3.10 at this voltage. Comment on the results.

3.15 Show that the small-signal capacitance of a symmetrical diode in reverse bias is equal to that of a parallel-plate capacitor whose plates are separated by the total width of the depleted region in the junction.

3.16 By what factor do the depletion layer widths change in a symmetrical junction when the doping concentration is doubled (a) on both sides (b) on one side only of the junction?

3.17 Calculate the small signal *diffusion* capacitances of silicon diodes at $T = 300$ K having doping concentrations $N_D = 10^{23}\ \text{m}^{-3} \ll N_A$, when (a) I = 100 μA, $L_n = 1$ μm, (b) I = 10 mA, $L_n = 25$ μm. Assume $D_h = 10^{-3}\ \text{m}^2/\text{s}$ and $\tau_h = 20$ μs. Comment on the type of applications for which each diode might be suitable and unsuitable.

3.18 An n^+p diode with a forward voltage drop of 0.7 V has been passing a forward current of 25 mA for a long time, when it is suddenly connected to a reverse voltage source of 5 V having an output resistance of 50 Ω (see figure P3.1). Make a dimensioned sketch of the current waveform through the diode.
If the storage time $t_s = 10$ ns and recombination is negligible, estimate the transit time and the diode length.

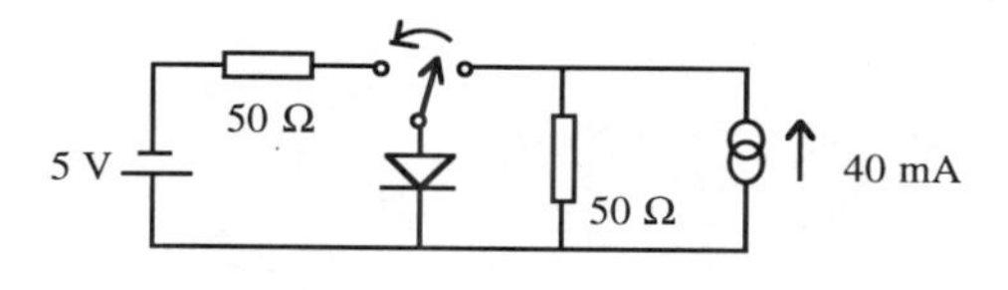

Fig. P3.1

3.19 Define the term *transit time* of minority carriers in an abrupt n^+p diode, and deduce an expression for it in terms of the diode's dimensions.
Why is the transit time likely to be reduced if the doping density is made to *decrease* with distance from the junction?
Estimate the electric field strength required to give a transit time equal to that of the diode in question 3.18, if diffusion is neglected. Calculate the built-in potential difference needed across the diffusion region to achieve this, and the ratio of doping concentrations at either end of the region which gives a built-in potential difference of this value.

3.20 After the diode in question 3.18 has been off for a long time, the switch is thrown back to turn it on. Sketch the diode current waveform, and explain its shape.

3.21 A diode with a transit time τ_t is connected as shown in Fig. P3.2, and the circuit is subjected to a voltage step of magnitude V_{in}. Model the diode as an *R–C* combination (i.e. assume that a large-signal model with appropriate fixed values of *R* & *C* is valid), and solve the circuit equations to show that the diode voltage undergoes a transient with time-constant $\frac{(C_1+C)}{1/R_1+1/R}$, and magnitude $V_{in}\left\{\frac{C_1}{C+C_1} - -\frac{R}{R+R_1}\right\}$. Hence show that the condition for the absence of a transient is that $R_1C_1 = \tau_t$.

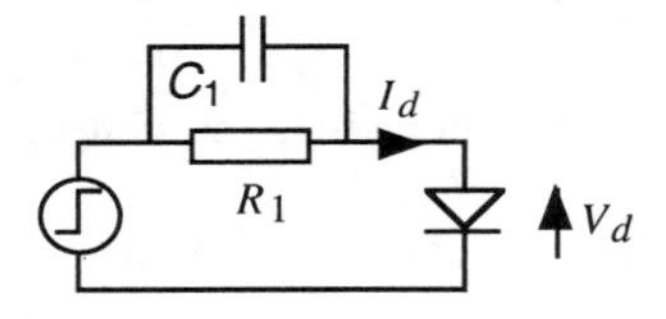

Fig. P3.2

Chapter 4

MOSFETs

4.1 Construction

The origin of the field effect transistor (FET) goes back as far as 1926. The basic idea was that it might be possible to make a 'voltage-controlled resistor' by varying the resistance between two contacts on the surface of a semiconductor (the SOURCE and the DRAIN contacts) with the aid of a third electrode called the GATE. Varying the gate voltage changes either the

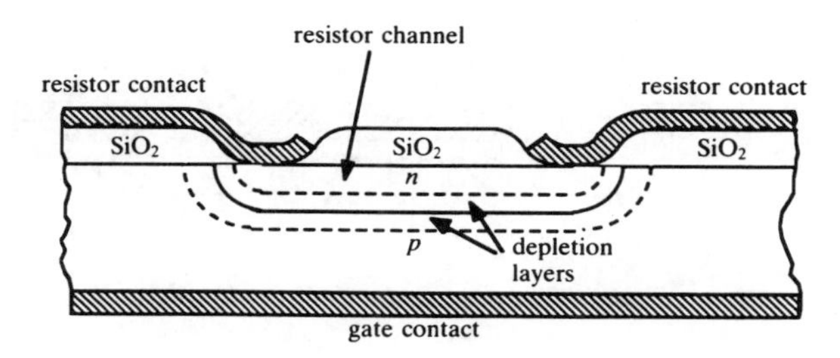

Fig. 4.1 A resistor consisting of an n-type region at the surface of a p-type wafer (shown in section) could be varied in value by widening or narrowing the depletion layer at the p–n junction, through changes in the voltage at the gate contact.

number of current carriers in the resistor, or the resistor thickness. Recalling how a resistor can be made by diffusing, for example, an *n*-type channel into a *p*-type substrate, as in Fig. 4.1, we could envisage changing the width of the depletion layer at the *p–n* junction with a variable bias voltage between *n*-type and *p*-type regions, so that the thickness of the conducting channel is varied. This is the principle behind the operation of the JUNCTION FET, abbreviated as JFET, though its construction normally differs somewhat from that shown in Fig. 4.1.

By varying the concentration of carriers in the conducting channel — and only incidentally its thickness — another kind of FET is produced. When this is made as a sandwich structure of metal/oxide/semiconductor, it is known as a Metal-Oxide-Semiconductor Field Effect Transistor, abbreviated as either MOSFET or MOST. In this chapter, we shall concentrate exclusively on the MOSFET, as it is currently so much more widely used than the JFET.

In the MOSFET, the *p–n* junction of the JFET is replaced by a thin insulator (silicon dioxide, SiO_2) backed by a metallic electrode called the gate. The usual construction is shown in cross section in Fig. 4.2. The device illustrated is constructed at the surface of a *p*-type silicon wafer called the SUBSTRATE. In the next section we explain how, by suitably biasing the gate electrode with a positive voltage relative to the semiconductor substrate, free electrons are attracted to the surface of the semiconductor immediately below the thin SiO_2 insulator. These electrons form a thin conducting layer called the CHANNEL between the two *n*-type doped regions which are labelled source and drain in Fig. 4.2. The conductance between source and drain is thus enhanced by the positive voltage applied to the

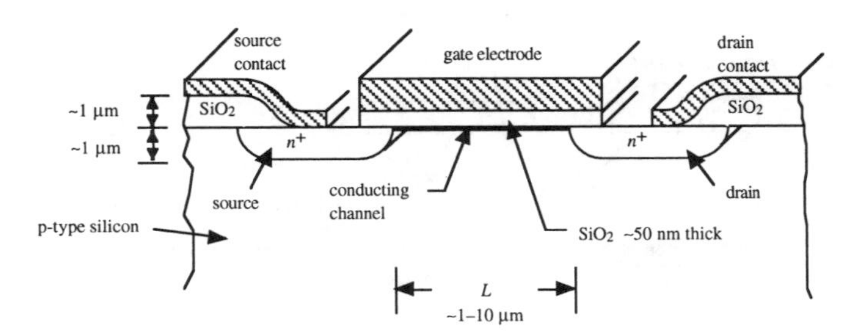

Fig. 4.2 The construction of a Metal-Oxide-Semiconductor Field Effect Transistor, shown in cross-section.

gate, provided the voltage is larger than a certain threshold value. Beyond this threshold voltage, we shall see that an increase in the gate voltage causes an increase in the *electron concentration* in the channel. This action contrasts with that in the JFET, where changes in channel resistance are a result of changes in channel depth, not carrier concentration.

In the absence of a positive voltage on the gate electrode, only a negligible current can flow between the *n*-type source and drain if a voltage is applied between them, since the region between them is then *p*-type, and one of the two *p–n* junctions must be reverse biassed by the source-drain voltage. The transistor is then off.

The transistor illustrated in Fig. 4.2 is known as an *n*-CHANNEL ENHANCEMENT FET, to distinguish it from three other types of MOSFET. If the source and drain regions are made by inserting *p*-type dopants into the surface of an *n*-type silicon wafer, a *p*-channel enhancement FET is formed. Depletion-mode FETs with either type of channel, in which the conducting channel exists even when the gate electrode is not biassed, will be discussed later.

4.2 Action of the insulated gate

To understand the action of the gate, we need to explain how the semiconductor surface, where the conducting channel forms, can apparently change its character from *p*-type to *n*-type, as a consequence of a change in gate potential — we say that the surface becomes INVERTED. This can only happen if there is a potential difference between the surface of the semiconductor and the *p*-type silicon deep in the bulk of the wafer. We shall ignore the presence of the source and drain terminals in this section, and consider them later.

It is instructive to compare the situation in a MOSFET having a bias voltage on its gate to the junction region in a *p–n* diode, for in the latter the concentration of electrons and holes at a point within the depletion region is connected with the potential there. The potential controls the local value of the energy difference $(E_c - E_F)$ between conduction band and Fermi level, which in turn affects the electron concentration, through the equation:

$$n = N_c \exp[-(E_c - E_F)/kT]$$

In the unbiased MOSFET, $(E_c - E_F)$ is large everywhere, as illustrated in Fig. 4.3(a).

A small positive potential applied to the gate, relative to the substrate, repels holes from the surface of the semiconductor immediately below the gate and creates a depletion layer there. An increase in the gate potential can result in enough potential difference appearing across this depletion layer to cause the energy difference $(E_c - E_F)$ to become quite small at the surface of the semiconductor. This situation is pictured in the energy level diagram of Fig. 4.3(b). At the surface the concentration, n, of electrons is much greater than the hole concentration there, and the nature of the surface is thus "inverted". The mobile charges in the channel are illustrated in Fig. 4.3(c), as well as the fixed negative charges on the acceptor ions beneath the channel in the depletion layer. The positive charge shown on the gate equals the sum of the negative charges, the structure being neutral as a whole.

Panel 4.1 explains how the gate acts in rather more detail, by pursuing the analogy with a *p–n* junction. If desired, this explanation can be omitted at a first reading.

If the gate voltage is increased further, the additional positive charge placed on the gate is now balanced by an equal increase of negative charge in the channel, increasing the electron concentration, and with it the channel conductivity. For this to happen, the conduction band edge E_c in the channel region need only move a little closer to the Fermi level E_F. For example, a change in $(E_c - E_F)$ of about 60 mV (2.4kT) is needed to produce an increase of $e^{2.4}$ (= 10) in electron concentration. As a consequence, *the voltage drop across the depletion layer beneath the channel is almost unaltered by the additional gate voltage, and so the total charge in the depletion layer remains virtually unchanged.* This important result will be useful in the next section when we calculate the current between the drain and source of the transistor.

In summary, there is a more or less well-defined voltage at which the channel becomes conducting. It is called the THRESHOLD VOLTAGE, and is typically in the range 0.5–3 V. For gate voltages below the threshold voltage, the gate induces changes in the width of the depletion layer, and the channel conductance is negligible. Above the threshold, the gate voltage induces changes in the electron concentration in the conducting channel, increasing its conductivity as the voltage rises.

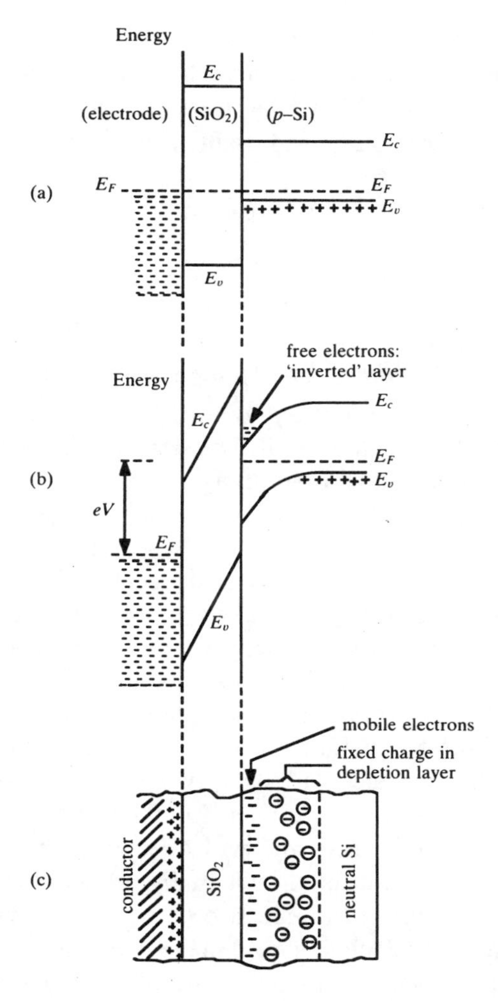

Fig. 4.3 (a) Energy band diagram at zero bias voltage of an ideal gate-insulator-semiconductor sandwich. Conduction and valence bands in the SiO_2 are also shown. (b) The same, with a gate bias voltage sufficient to "invert" the semiconductor surface, and (c) distribution of the charges when the channel is conducting.

The threshold voltage itself is normally defined as that gate voltage which just results in a surface electron concentration equal to the concentration of majority carriers in the substrate, far from the transistor. Discussion of the magnitude of the threshold voltage and what controls it are postponed to section 4.10. In the next section, the current which flows in the channel when the source and drain are connected to a voltage source will be calculated.

PANEL 4.1

Effects of the insulated gate on the semiconductor energy levels

The energy band diagram of a *p–n* junction is repeated in Fig. 4.4(a) for comparison with the MOSFET. The slope of the energy levels at the centre of the *p–n* junction represents a potential gradient; a high electric field strength $E = -dV/dx$ exists there. We can establish the same potential gradient on the *p*-type side of the junction by replacing the *n*-type silicon with an externally biased electrode (the gate), separated from the *p*-type region by an insulator as in Fig. 4.4(b). The voltage applied by the source is divided between the insulator and the depletion layer, since they are in series with one another. By suitably choosing the value of the applied voltage, the potential gradient just inside the silicon, at the interface with the SiO_2, can be adjusted to the value in Fig. 4.4(a), at the interface with the *n*-type silicon. Since in both cases no net current flows across the interface, the rest of the *p*-type silicon cannot know the difference, and will have an identical energy band diagram in both cases, as shown in Fig. 4.4(c), where the SiO_2 conduction and valence bands are also shown. Thus a suitable external voltage has induced a depletion layer in the *p*-type silicon. At the surface of the silicon, the electron and hole concentrations have their intrinsic values because the Fermi level E_F lies midway between E_c and E_v.

Note that the net negative charge in the depletion layer (on the ionised acceptor atoms) is exactly balanced by an equal positive charge on the external electrode, as illustrated in Fig. 4.4(d).

(*Continued*)

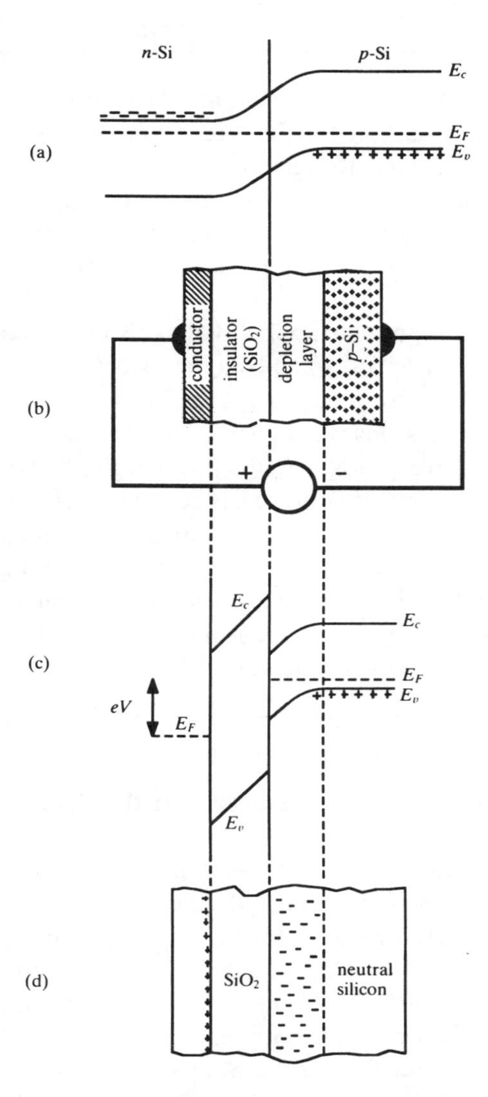

Fig. 4.4 (a) Energy band diagram of a *p–n* junction (b) replacement of half of the junction by an insulator and conductor, showing only the mobile holes in the semiconductor (c) energy band diagram of the arrangement in (b), and (d) distribution of net charge across the same.

PANEL 4.1 (*Continued*)

Now consider the effect of changing the magnitude of the voltage source shown in Fig. 4.4(b). If it is removed entirely — and if there is no contact potential difference between the *p*-type silicon and the external electrode* — we end up with the energy level diagram in Fig. 4.3(a). There is no depletion layer, and the free carrier concentrations are the same everywhere within the semiconductor, i.e. the surface is *p*-type.

If, on the other hand, the applied voltage were increased to *double* its previous value, the energy band diagram would look like Fig. 4.3(b). Now the conduction band edge E_c is pulled down towards E_F: at the SiO_2-Si interface, it is close enough to E_F for the electron concentration, being proportional to $\exp[-(E_c - E_F)/kT]$, to be high. The hole concentration is correspondingly low: the surface has become *n*-type. The thin surface layer has become 'inverted' by the applied gate voltage, and forms a conducting channel which is isolated from the underlying *p*-type silicon by the depleted layer between them. The mobile charge in the channel is illustrated in Fig. 4.3(c), as well as the fixed negative charges on the acceptor ions beneath the channel, in the depletion layer.

4.3 Calculation of the channel charge and the drain current

Figure 4.5 shows the *n*-channel MOSFET in cross-section, with voltage sources connected as for a normal operation. The voltages are all conventionally measured relative to the source terminal, which is usually connected directly to a contact on the bulk of the semiconductor called either the BULK or SUBSTRATE terminal. The notation for these voltages is as given in the figure. For example, the symbol V_{DS} refers to the the potential at the Drain relative to the Source, while V_{GD} would denote the gate potential relative to the drain. Note also that the drain is always, by definition, positive with respect to the source in an *n*-channel transistor.

*The requirement for zero contact potential difference can be (almost) achieved by making the gate electrode of *p*-type silicon, which is so heavily doped as to be effectively metallic. By this, we mean that the Fermi level E_F lies within the valence band, just below E_v.

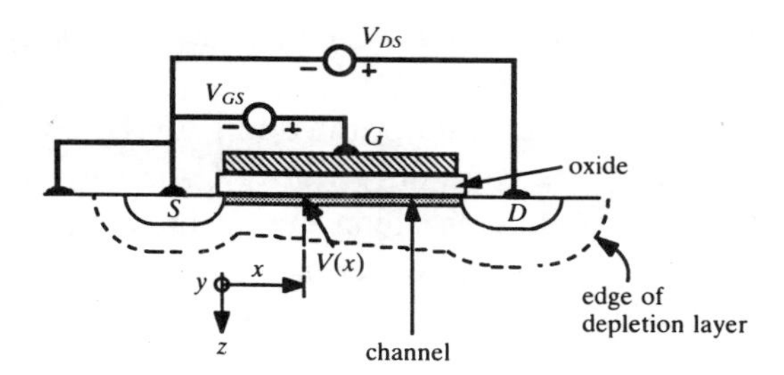

Fig. 4.5 Section through a MOSFET, showing voltages applied in normal operation.

We wish to find a relationship between the source-drain current and the applied voltages. The conductance of the channel connecting the source and the drain depends on the concentration of electrons there. This in turn is directly related to the charge density Q_c that these electrons carry, expressed per unit area of the x–y plane in the figure, i.e. per unit area of the channel itself. So the first task is to find a suitable expression for the channel charge density in terms of the voltages applied to the transistor.

We assume that $Q_c = 0$ for all values of V_{GS} up to and including the threshold voltage, for which the symbol V_T will be used. Using the principle of superposition, we can discuss *changes* in Q_c and V_{GS} above threshold independently of the charges and voltages present *at* threshold, provided only that the latter are fixed. We can therefore relate the excess of V_{GS} above its threshold value, i.e. $(V_{GS} - V_T)$ to the channel charge density Q_c which accompanies that rise.

Now the excess voltage $(V_{GS} - V_T)$ is entirely dropped across the oxide, since almost no change takes place in the depletion layer below it once the threshold is exceeded, as was established in section 4.2. Hence, we can treat the sandwich formed by the oxide layer, the negative charge Q_c in the channel and the (positive) counter-charge on the gate, as if it were a parallel-plate capacitor. Thus we can relate charge and voltage by the capacitance:

$$Q_c = -C_{ox}(V_{GS} - V_T)$$

Since Q_c is expressed per unit area, the capacitance C_{ox} is the capacitance *per unit area* of the oxide sandwich referred to. Its value is therefore equal to ε_{ox}/t_{ox}, i.e. the oxide's permittivity divided by its thickness. C_{ox} is usually referred to as the OXIDE CAPACITANCE, and is typically between 10^{-4} and 10^{-3} pF/μm^2.

So far, we have taken no account of the drain-source voltage V_{DS}, so that the above equation for Q_c is valid only when V_{DS} is zero. Before going on to relate Q_c to the current flowing along the channel, we have to take into account the fact that the potential difference across the oxide sandwich is lower at the drain end of the channel than at the source. This is because the drain voltage is always positive with respect to the source. As remarked above, the drain is *defined* as the more positive of the two terminals.

At the drain end of the channel, the potential relative to the source is V_{DS}. There can be no free carriers in the channel at this point, unless the voltage across the oxide sandwich, which here is the gate-drain voltage V_{GD}, exceeds the threshold voltage V_T. The charge Q_c per unit area at the drain end of the channel is determined by the value of $(V_{GD} - V_T)$, while at the source it is proportional to $(V_{GS} - V_T)$.

Now the voltage across the gate-oxide-channel sandwich does not change abruptly to V_{GD} at the drain, but changes steadily from the value V_{GS} to V_{GD} along the length of the channel, as illustrated in Fig. 4.6.

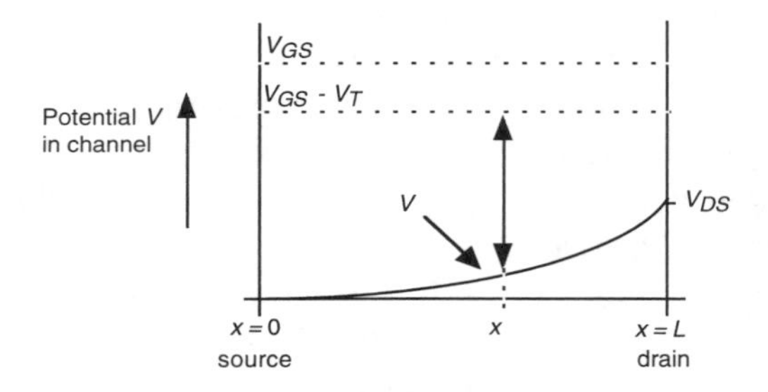

Fig. 4.6 The voltage V of a point in the channel, relative to the source, changes gradually along the channel. The double-headed arrow shows the magnitude of the net voltage which creates the charge density Q_c in the channel.

At some intermediate point in the channel which lies at a distance x from the source, let the voltage relative to the source be V. The voltage across the oxide sandwich is then $(V_{GS} - V)$. The charge per unit area in the channel at that point is therefore just

$$Q_c = -C_{ox}(V_{GS} - V - V_T) \tag{4.1}$$

You should note that this equation is correct only when $(V_{GS} - V)$ exceeds V_T, otherwise $Q_c = 0$. In other words, the channel charge in this MOSFET can never be positive. Equation (4.1) replaces the previous equation for Q_c, which applies only when $V_{DS} = 0$.

Now the potential gradient dV/dx along the channel, which is illustrated in Fig. 4.6, causes the electrons to move towards the drain with a velocity given by:

$$v = \mu_e \frac{dV}{dx} \tag{4.2}$$

The current is the total amount of charge passing a point at any position x in the channel in unit time, which is just $Q_c W v$, where W is the channel width measured along the y axis in Fig. 4.5. Hence the current I_D from drain to source is

$$I_D = -Q_c W v = C_{ox}(V_{GS} - V - V_T) W \mu_e \frac{dV}{dx} \tag{4.3}$$

where we have used both equations (4.1) and (4.2). The current I_D is called the DRAIN CURRENT, and is conventionally taken to be positive when flowing *into* the drain terminal.

Since I_D is constant and independent of x, eqn. (4.3) can be regarded as a differential equation in $V(x)$ which can be solved by by integrating it directly. We shall avoid the need for finding the function $V(x)$ by integrating the equation with respect to x, between the limits $x = 0$ and $x = L$, i.e. between the source and the drain. Thus, after a change of variable on the right hand side, the integral becomes

$$\int_0^L I_D dx = \int_0^{V_{DS}} W C_{ox} \mu_e (V_{GS} - V - V_T) dV$$

Since I_D and μ_e are both independent of x, we find after integrating and dividing the equation by L:

$$I_D = \frac{W}{L} C_{ox} \mu_e \left[(V_{GS} - V_T) V_{DS} - \frac{V_{DS}^2}{2} \right] \tag{4.4}$$

which is the final result, valid when $V_{DS} < V_{GS} - V_T$.

This equation, relating I_D and V_{DS}, describes the shape of the so-called output characteristic of the transistor, and is plotted for several V_{GS} values in Fig. 4.7(a). When V_{DS} is small, each curve is nearly linear, with a slope which is proportional to $(V_{GS} - V_T)$ — the behaviour of a 'voltage-controlled resistor' which we were seeking in section 4.1. Note that eqn. (4.4) is also correct for negative values of V_{DS}, although we normally refer to the more positive output terminal of the transistor as the drain. The extension of each characteristic in Fig. 4.7(a) into the third quadrant is nevertheless shown there, and is a continuation of each parabola, so that each curve is not symmetrical with respect to the origin.

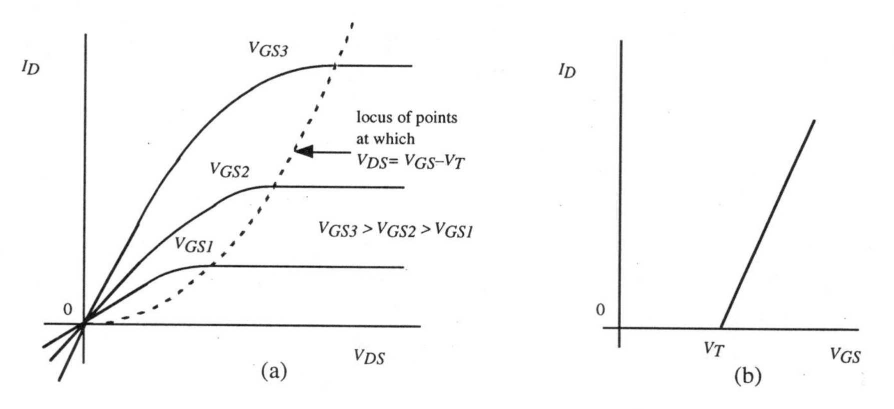

Fig. 4.7 (a) Ideal output characteristics of an *n*-channel MOSFET (b) the corresponding drain current-gate voltage transfer characteristic.

At higher positive values of V_{DS}, the slope of each of the characteristics in Fig. 4.7(a) falls, until it reaches zero when $V_{DS} = V_{GS} - V_T$. This is easily proved by differentiating eqn. (4.4) with respect to V_{DS}. Note that at this point, where $\partial I_D/\partial V_{DS}$ just becomes zero, the gate-drain voltage V_{GD} is just equal to the threshold voltage: the channel charge is just about to disappear at the drain end (see eqn. (4.1)). Equation (4.4) is not valid beyond this point: the drain current actually remains almost constant (we say that it becomes saturated) as V_{DS} increases further, as will be explained in section 4.4.

Operating regions of the characteristics

The region of Fig. 4.7(a) to the left of the dashed line, in which eqn. (4.4) remains valid, is often known as the TRIODE REGION of the characteristics. In this book we shall refer to it for clarity as the UNSATURATED REGION, to distinguish it clearly from the SATURATION REGION, where the drain current does not increase with V_{DS}. The initial portion of the the unsaturated characteristic, where $V_{DS} \ll V_{GS} - V_T$, is sometimes called the LINEAR region of operation.

A third principal region of the characteristics, in which the transistor is OFF, (i.e. no drain current flows), exists for all gate voltages below threshold. This can be seen in Fig. 4.7(b), which shows how the drain current I_D changes with gate-source voltage V_{GS} when V_{DS} is held constant at a small value. One practical point to note about this diagram is that the threshold of conduction is not as sharp as the above discussion would imply. The threshold voltage is normally defined for a real device by extrapolating the linear part of this characteristic back to the voltage axis. The reason for the curvature near this point is covered in Panel 4.2, which should be read following the discussion of the saturation region of the characteristics in the next section.

4.4 The saturation region of the output characteristic

When $V_{DS} \geq V_{GS} - V_T$ the transistor is said to be SATURATED, meaning that the drain current no longer rises strongly with V_{DS}. At the onset of saturation, the potential relative to the source varies along the channel, as shown in Fig. 4.8(a), reaching the value $(V_{GS} - V_T)$ at the drain. At this point, the voltage between gate and drain, V_{GD}, just equals the threshold voltage V_T. The channel is said to be PINCHED OFF at the drain, and the drain-source voltage V_{DS} at which this occurs is often referred to as the pinch-off voltage. Note, however, that the pinch-off voltage does not have a fixed value, as it is equal to $(V_{GS} - V_T)$.

Any further increase in V_{DS} ensures that the channel becomes fully depleted of carriers at the end nearest the drain. Thus, an extremely short depletion region forms between the channel itself and the drain, as illustrated in Fig. 4.8(b). The electrons in the channel must flow across this depleted region in order to reach the drain. They can do this with the help of even

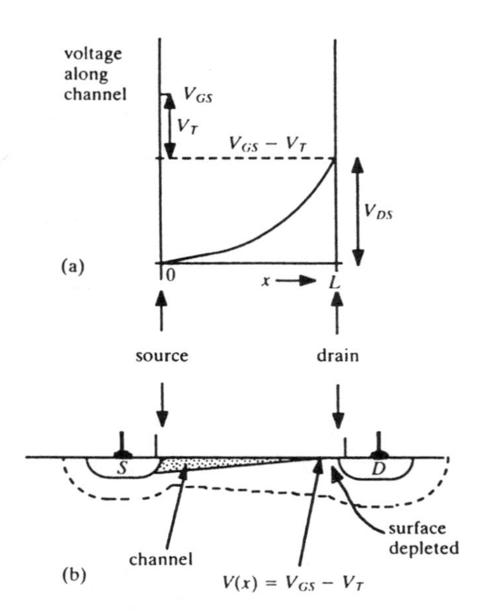

Fig. 4.8 (a) Voltage variation along the channel in saturation (b) a depletion region of length L_D is formed in saturation at the drain end of the channel.

a small voltage difference, because the distance involved is so small that a large electric field strength is easily created. At the point where the conducting channel ends (is 'pinched-off'), the potential is equal to the pinch-off voltage $(V_{GS} - V_T)$. Note that the difference between the voltage V_{DS} and the pinch-off value of $(V_{GS} - V_T)$ appears across the short, high-resistance depletion layer, and is enough to give the few electrons present there a high velocity towards the drain. Indeed, they travel at their maximum speed, the *saturation velocity*.

We expect the current to rise no further, because an increase in V_{DS} no longer raises the velocity of carriers in the conducting channel itself, but merely lengthens the depleted region slightly. The drain current thus saturates, and the characteristics are shown horizontal on the theoretical plots in Fig. 4.7(a). The drain current in saturation is simply found by putting $V_{DS} = (V_{GS} - V_T)$ into eqn. (4.4), thus

$$I_D = \frac{1}{2}\mu_e C_{ox}\frac{W}{L}(V_{GS} - V_T)^2 \quad \text{in saturation} \tag{4.5}$$

However, constancy of the saturation current depends upon maintaining a constant channel length. In practice, the depletion layer at the drain end lengthens as the voltage across it increases. The channel length L decreases slightly in consequence, and the drain current rises a little. This is known as the EARLY effect, after the first person to describe the rather similar behaviour observed in bipolar transistors. It is also referred to as CHANNEL LENGTH MODULATION, and it will be explained at greater length in section 4.6. Figure 4.9 shows some transistor characteristics where the effect is clearly seen.

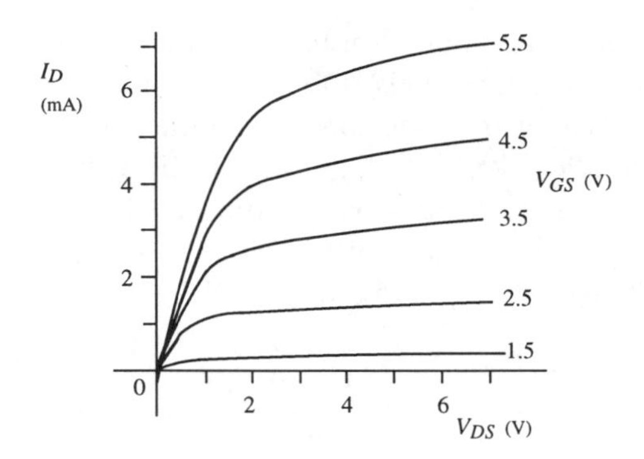

Fig. 4.9 Transistor characteristics showing the effect of channel length modulation in the saturation region.

We can use eqn. (4.5) to find the boundary between the unsaturated and saturation regions on Fig. 4.7(a) by naming the values of I_D and V_{DS} at the onset of saturation as I_{DSat} and V_{DSat} respectively. Then, since $V_{DSat} = V_{GS} - V_T$, eqn. (4.5) gives

$$I_{DSat} = \frac{1}{2}\mu_e C_{ox} \frac{W}{L} V_{DSat}^2$$

This is the equation of a parabola, which is shown dashed in Fig. 4.7(a).

It is worth noting that the term saturation has a quite different meaning in the case of the bipolar transistor, as will be explained in chapter 5.

PANEL 4.2

The pinched-off channel: subthreshold conduction

When $V_{DS} \geq V_{GS} - V_T$, i.e. when the *drain* end of the channel is pinched off, the carrier concentration there is not zero but merely small in magnitude, just as in the depletion region of a *p–n* junction. The gate voltage is unable to maintain *strong inversion* at the drain end of the channel; we say that it supports *weak inversion*. Similarly, weak inversion occurs at the source if we make V_{GS} slightly less than V_T: the transistor is off and, in the normal way, we would then say that the transistor is *below threshold*.

The term 'weak inversion' implies that the channel region *does* contain some free electrons. This is the so-called sub-threshold region of operation, in which the electron concentration in the channel is sufficient only to allow very small currents to flow. In this regime, the following approximate expression for the drain current I_D applies, provided that V_{SB} and V_{DB} are not too close to the pinch-off voltage $V_{GS} - V_T$, which is abbreviated to the symbol V_p:

$$I_D = 2\mu C_{ox}\frac{W}{L}\left(\frac{kT}{e}\right)^2\left\{\exp\left(\frac{V_p - V_{SB}}{kT/e}\right) - \exp\left(\frac{V_p - V_{DB}}{kT/e}\right)\right\}$$

Here the source and drain voltages are expressed relative to the substrate, or bulk silicon.

More accurate (but still relatively simple) equations for I_D are available for use when V_{SB} and V_{DB} are close to V_p.

The sub-threshold region is important commercially for two reasons:

(i) it determines the 'off' current of a transistor in many digital circuits, and hence the quiescent current drawn by each of the millions of switches used in a complete CMOS integrated circuit,
and

(ii) it can be used instead of the unsaturated region for low-power applications (e.g. in watches, satellites and body implants), and is likely to become more important due to a growing appreciation by more engineers of the circuit design techniques needed for its use.

4.5 The small-signal equivalent circuit

As with any electronic device, an equivalent circuit can be constructed to model the way small changes in input voltage affect both the input current and the output voltages and currents. It is used to represent the behaviour of the transistor when it is employed in a circuit designed to amplify small currents or voltages at the input terminals. The simplest equivalent circuit for a MOSFET is shown in Fig. 4.10(a), alongside the symbol of the transistor which it models. Note the arrow pointing from the *p*-type substrate to the *n*-channel, which identifies the channel type.

The convention followed in chapter 3 for the use of small letters with small subscripts for small-signal voltages and currents is continued here. The circuit of Fig. 4.10(a) is used to model the transistor in the connection shown in Fig. 4.10(b). This is known as the common-source connection, since the source terminal is common to both the input and output.

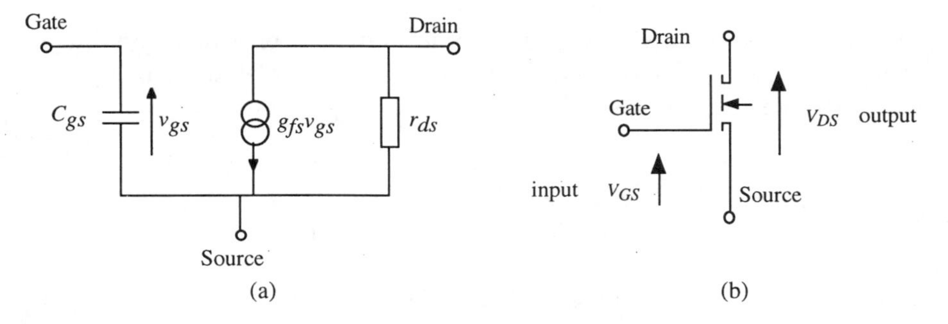

Fig. 4.10 (a) Small-signal equivalent circuit of a MOSFET in the common-source connection (b) Circuit symbol of the *n*-channel MOSFET, shown in the common-source connection.

It is necessary to treat the unsaturated and saturation operating regions of the transistor separately, as in earlier sections. Within each region of operation, the component values depend upon the bias voltages, i.e. upon the dc components of the voltages V_{GS} and V_{DS}.

When used in an amplifier, a MOSFET is usually biased in its saturation region, so we shall concentrate more upon the saturation region in what follows.

Let us first consider how to define each of the component values in Fig. 4.10(a).

The gate-source input capacitance C_{gs}

A MOSFET has almost infinite input resistance, owing to the excellent insulating properties of the gate oxide. Hence, we normally need only to take account of the input capacitance.

The input capacitance can be defined as the small-signal charge q_g on the gate electrode divided by the small-signal gate-source voltage v_{gs}. Thus

$$C_{gs} = \frac{q_g}{v_{gs}} = \frac{\partial Q_G}{\partial V_{GS}}$$

The transconductance g_{fs}

The dependence of the small-signal drain current i_d on small-signal gate-source voltage v_{gs} is modelled by a current generator of strength $g_{fs}v_{gs}$ in Fig. 4.10, where g_{fs} is given by

$$g_{fs} = \frac{i_d}{v_{gs}} = \left(\frac{\partial I_D}{\partial V_{GS}}\right)_{V_{DS}=\text{const}}$$

g_{fs} is referred to as the forward TRANSCONDUCTANCE* of the transistor in the common-source connection. It has the dimensions of a conductance, but is unlike an ordinary conductance in that it connects an *output* current with an *input* voltage. The two subscripts *f* and *s* are used to distinguish this transconductance from others which are sometimes needed, for example the reverse transconductance g_{rs}, used when the drain and source are interchanged, and similar transconductances defined for cases when the terminal which is common to both input and output, is not the source (denoted by the second subscript *s* to the symbol), but the gate or drain.

*The symbol g_m is often used in place of g_{fs}, which is the internationally agreed symbol for this component.

Note from the equation that in evaluating the transconductance, V_{DS} is to be held constant, i.e. $v_{ds} = 0$. Otherwise, some of the small-signal current generated would flow in the resistance r_{ds} in the figure. If we are to measure the strength of the current source in Fig. 4.10 by measuring the external drain current, it must be measured in a short-circuit across the terminals.

The output resistance r_{ds}

The resistance r_{ds} in Fig. 4.10 is commonly known as the OUTPUT RESISTANCE of the transistor, since the drain and source are the output terminals of the amplifier. It models that part of the change in drain-source voltage V_{DS} which is proportional to I_D. Thus the definition of r_{ds} is

$$r_{ds} = \frac{v_{ds}}{i_d} = \left(\frac{\partial V_{DS}}{\partial I_D} \right)_{V_{GS}=\text{const}}$$

We shall use these definitions to find expressions for the component values in the unsaturated region of operation, since this is the easiest case to consider. The saturation region requires a little more development of the ideas introduced in earlier sections.

4.6 Component values in the equivalent circuit: the unsaturated region of operation

The gate-source input capacitance

The two input terminals, the gate and the source, present a very high d.c. input resistance, typically above $10^{12}\ \Omega$ and a small capacitance C_{gs} in the range 0.01–10 pF, which is shown in Fig. 4.10. The input resistance is sufficiently high that for practical purposes it is commonly omitted, as here. Hence at all useful frequencies above d.c., the input impedance is

$$Z_{in} = (j\omega C_{gs})^{-1}$$

Since C_{gs} is small, the input impedance has a very large magnitude, and indeed may be assumed infinite for practical purposes, except at high frequencies, in all regions of operation. We shall discuss what determines the value of C_{gs} later, along with other capacitances which play a role at high frequencies.

Transconductance in the unsaturated region of operation

The definition of g_{fs} above is easily used in conjunction with eqn. (4.4) for I_D to find an expression for the transconductance in the unsaturated region. Differentiating eqn. (4.4) gives the result

$$g_{fs} = \mu_e C_{ox}\left(\frac{W}{L}\right)V_{DS} \quad \text{(unsaturated)}$$

Output resistance in the unsaturated region

By differentiating eqn. (4.4) with respect to V_{DS} it is easy to show that

$$r_{ds} = \mu_e C_{ox}\left(\frac{W}{L}\right)[V_{GS} - V_T - V_{DS}] \quad \text{(unsaturated)}$$

4.7 Component values in the equivalent circuit: the saturation region of operation

Input capacitance

The d.c. input resistance remains effectively infinite.

The input impedance is identical to that given for the unsaturated region, thus:

$$Z_{in} = (j\omega C_{gs})^{-1}$$

although the value of C_{gs} differs somewhat from that in the unsaturated region, as will be discussed later.

Transconductance in the saturation region

To evaluate the transconductance in the MOSFET's saturation region, we differentiate eqn. (4.5) with respect to V_{GS}, and find that

$$g_{fs} = \mu_e C_{ox} \frac{W}{L}(V_{GS} - V_T) \quad \text{(in saturation)} \tag{4.6}$$

Here, V_{GS} denotes the mean gate-source voltage difference (i.e. the d.c. or bias component). Unless the input voltage v_{gs} is small compared to V_{GS}, the transconductance g_{fs} varies with signal amplitude, thereby creating a distortion of the collector current waveform. Note that the designer can vary g_{fs} at will by choosing the ratio W/L.

To estimate a typical value of g_{fs}, we first estimate C_{ox}. The relative permittivity ε_{ox} of SiO_2 is close to 4.0. Thus, assuming an oxide thickness t_{ox} of 0.05 μm, we have

$$C_{ox} = \frac{\varepsilon_{ox}\varepsilon_o}{t_{ox}} = \frac{4 \times 8.8 \times 10^{-12}}{5 \times 10^{-8}} = 7.0 \times 10^{-4} \ \text{pF}/\mu\text{m}^2 .$$

The next point to note is that the mobility μ_e of electrons is about one half of the value used in Chapter 1 appropriate to the doping concentration in the channel, because of extra collisions of the electrons with the surface of the Si at the interface with the SiO_2. The channel thickness is comparable to the mean free path, so that such collisions are relatively frequent. For example, when the doping is low, μ_e is reduced to about 0.07 m^2/Vs. Hence when $W/L = 10$, we have typically

$$\begin{aligned} g_{fs} &= 0.07 \times 7.0 \times 10^{-4} \times 10(V_{GS} - V_T) \quad \text{(in saturation)} \\ &= 0.28 \text{ mS for every volt by which } V_{GS} \text{ exceeds } V_T . \end{aligned}$$

When compared with the transconductance of a bipolar transistor evaluated in Chapter 5, this is a rather low value when voltages of just a few volts are used.

Output Resistance in the saturation region (The Early effect, or channel length modulation)

The output resistance cannot be calculated directly from eqn. (4.5) for the drain current, since V_{DS} does not appear explicitly there. However, as already explained, the drain current does increase slightly with V_{DS}.

As we have seen, when the gate-drain voltage V_{GD} falls below the value of the threshold voltage V_T, the drain current saturates because the channel at the drain end becomes depleted of carriers. As the drain voltage rises further, with V_{GS} constant, the point in the channel at which pinch-off begins moves slightly away from the drain, as shown in Fig. 4.11.

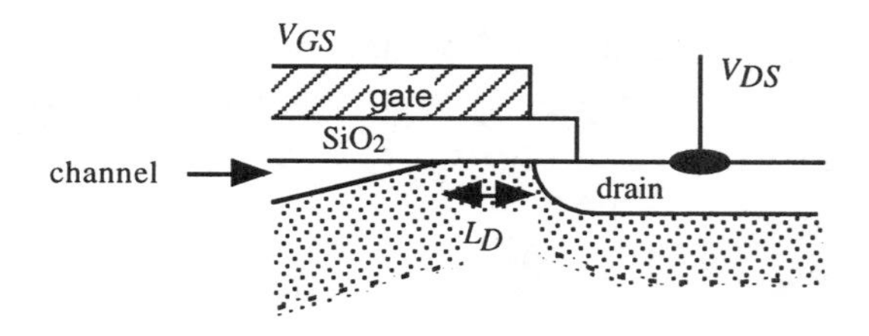

Fig. 4.11 An enlarged cross-sectional view of the channel at the drain end where, in saturation, the channel is pinched-off (i.e. depleted) over a length L_D.

This reduces the effective channel length of the depleted region by the length L_D, so that the drain current may be found by modifying eqn. (4.5) as follows

$$I_D = \frac{1}{2}\frac{\mu C_{ox} W}{(L - L_D)}(V_{GS} - V_T)^2 \tag{4.7}$$

As L_D rises with increasing V_{DS}, I_D also rises slightly, giving the transistor its finite output resistance r_{ds}. The dependence of L_D upon V_{DS} is not easily calculated, however, and approximations are usually used.

The simplest model of this effect, which is provided within the SPICE computer modelling package, assumes that $L_D = \Lambda V_{DS}$, where Λ is an adjustable parameter chosen by the user, independent of transistor voltages or currents. The corresponding value of r_{ds}, the small-signal output resistance is then simply found by differentiating eqn. (4.7) to give

$$r_{ds} = \left(\frac{\partial I_D}{\partial V_{DS}}\right)^{-1} = \frac{(L - L_D)}{\Lambda I_D} = \frac{V_A}{I_D}$$

where $V_A = (L - L_D)/\Lambda$ is called the EARLY VOLTAGE. This parameter is approximately constant whenever $L \gg L_D$, i.e. in all except transistors with very short channels. Although there is no simple justification for the assumption that $L_D = \Lambda V_{DS}$, the simple expression for r_{ds} which results does indeed describe the behaviour of real transistors fairly well.

Since the voltage across the depleted channel length L_D is $V_{DS} - V_p$ (where V_p is the pinch-off voltage, equal to $V_{GS} - V_T$), then by applying conventional abrupt *p–n* junction theory, L_D might be expected to rise in proportion to $(V_{DS} - V_p)^{1/2}$. This is not often the case in practice, since the conventional theory applies only to the doping profile of a planar, abrupt junction, whereas the doping here usually varies with depth as well as distance along the channel. Most versions of SPICE usually provide the option of using this, more complicated, model for the Early effect, in spite of its limitations.

4.8 Small-signal capacitances and transient behaviour

Before discussing the small-signal circuit components which model the high-frequency behaviour of a MOSFET, it is helpful to consider the factors which limit the rate at which the drain current responds to a change in the gate-source voltage. The most fundamental of these is the so-called transit time.

The transit time

The maximum rate at which the drain current I_D can be switched by a change in input voltage is determined by the rate at which the charge in the channel is changed, because it is this charge which carries the current.

Suppose an *n*-channel enhancement MOSFET is initially its unsaturated (triode) region, with V_{DS} held constant. If V_{GS} is increased instantly, the mobile charge in the channel must rise in response eventually. Since thermal generation is slow, the extra carriers are supplied from the source. The time taken for these carriers to cross the channel (called the TRANSIT TIME) roughly represents the delay before the drain current reaches its new equilibrium value. An estimate of the shortest possible transit time is v_{sat}/L — the saturation drift velocity (referred to in section 1.8) divided by the channel length. This is about 10 picoseconds for a 1 μm long channel, since v_{sat} is approximately 10^5 m/s in silicon. The resulting delay is commonly negligible, compared with the time taken to charge the gate, as we now explain.

Charge control

In practice, the speed at which I_D changes is determined by two factors: (i) The rate at which the gate voltage can be changed (which is dependent on the R-C time constant of the gate circuit), and (ii) the rate at which the channel charge responds to changes in gate voltage, which is related to the transit time referred to above. Normally, the former dominates in all except the very fastest MOS integrated circuits, and the problem of how the charge redistributes itself within the transistor can be simplified by assuming that the gate voltage rises slowly enough to allow the charges to attain their steady-state values instantly. For example, in the SPICE simulation program, the static model is used to calculate the stored charge at each stage of the switching waveform. The rate at which V_{GS} rises is then determined by the charging current flowing into the gate via external resistances. It should be borne in mind that this approximation will break down if the gate voltage changes significantly within a timescale of the order of the transit time (e.g. within about 10 ps in a 1 μm channel).

The time constant for the gate circuit depends primarily on the internal resistance of the source which drives the gate, and on the various capacitances associated with the MOSFET. The three main charges stored in these capacitances are:

(i) the *total* charge Q_C in the channel*

---

*N.B. Q_C used here is not equal to Q_c, the charge per unit area used in eqn. (4.1), but is related to it, since $Q_C = \int_0^L Q_c W dx$.

(ii) the whole charge Q_{DEP} in the depleted bulk region (shown in Fig. 4.3(d)), and
(iii) the total charge Q_G on the gate.

Although the charges can be calculated directly, knowing the terminal voltages, it is often easier conceptually to think in terms of the *small-signal capacitances* between the terminals, remembering that these give only approximate results if used for large voltage swings. Further simplification results from taking into account only the three most important capacitances (referred to as the Meyer model), as shown in Fig. 4.12, in which we have shown the substrate or bulk (B) and the source (S), connected to one another, as assumed earlier. However, this is not always the case in real circuits, and we shall assume in this section that the potential of the substrate may differ from that of the source, i.e. $V_{SB} \neq 0$.

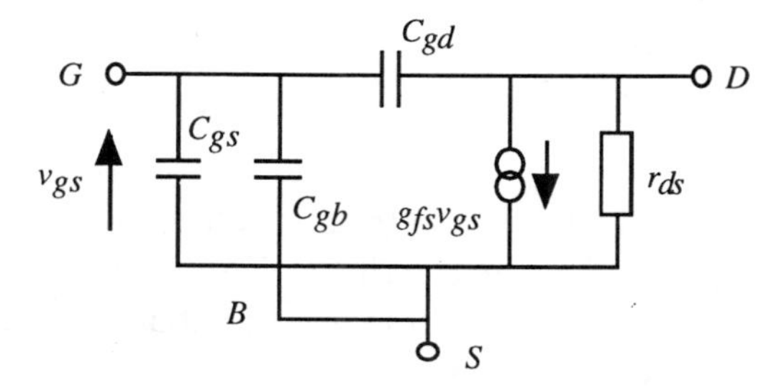

Fig. 4.12 The Meyer model for the small-signal equivalent circuit includes two capacitances omitted in the simplified circuit shown in Fig. 4.10.

The capacitances are defined in terms of the charges as follows:

$$C_{gs} = \frac{\partial Q_G}{\partial V_{GS}},$$

$$C_{gd} = \frac{\partial Q_G}{\partial V_{GD}},$$

$$\text{and} \quad C_{gb} = \frac{\partial Q_G}{\partial V_{GB}}$$

Each derivative is to be evaluated with the voltages across the other capacitances held constant, so that the change in charge on the gate is influenced only by the voltage of interest.

We consider first the so-called *intrinsic* capacitances, ignoring for the moment some small capacitances (to be discussed later) which result from the overlap of the gate electrode with both drain and source.

The three main regions of the transistor I–V characteristic must be considered separately. We shall consider an n-channel transistor in the following description, although a p-channel transistor behaves similarly when appropriate changes are made in the signs of voltages.

(a) *In the linear region* of operation (i.e. for values of drain voltage V_{DS} which are small compared to $V_{GS} - V_T$), the conducting channel is approximately an equipotential surface, and the electric flux lines from the gate charge to the channel can be thought of as being equally divided between source and drain. Then, as illustrated in Fig. 4.13(a),

$$C_{gs} = C_{gd} = \frac{1}{2} C_{ox} WL \quad (\text{when } V_{DS} \ll V_{GS} - V_T) . \tag{4.8}$$

To these values should be added the (small) values of the so-called *overlap capacitances*, C_{gso} and C_{gdo} respectively. These exist because the gate electrode itself must extend at least as far as the source and drain, otherwise the conducting channel would not reach the source or the drain and the transistor would fail to conduct current. C_{gso} and C_{gdo} are the capacitances between the respective electrodes in the absence of any free carriers in the conducting channel. In a transistor of symmetrical construction they have equal magnitudes.

In the whole of the unsaturated (triode) region of operation, C_{gb} is negligibly small, since changes in the substrate voltage cannot affect the gate charge, as long as V_{GS} is held constant. Changes in substrate voltage affect the charge Q_{DEP} in the depletion layer below the channel, but the sum of Q_C and Q_{DEP}, which equals $-Q_G$, is unaffected.

(b) In saturation (i.e. when $V_{DS} \geq V_{GS} - V_T$) the values of C_{gs} and C_{gd} differ; the presence of the high-resistance depleted region at the drain end of the channel ensures that the drain voltage does not significantly affect the amount of mobile charge in the channel. Hence C_{gd} drops to the value of the overlap capacitance C_{gdo} alone.

(a)

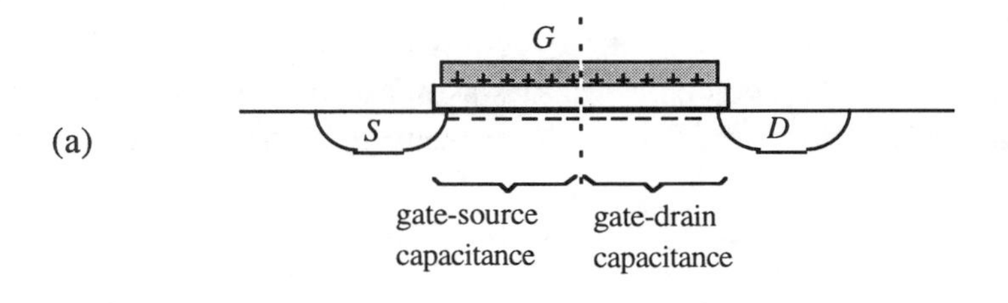

(b)

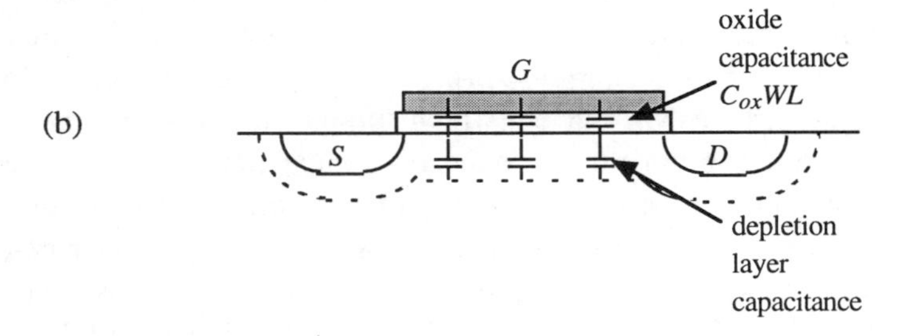

Fig. 4.13 Illustrating the division of the gate capacitance between source and drain in a MOSFET. (a) The gate capacitance is shared equally between source and drain when V_{DS} is small and V_{GS} exceeds the threshold, (b) below threshold, the depletion layer capacitance appears to be connected in series with the oxide capacitance, between the gate and the substrate.

Although C_{gs} might then be expected to equal $C_{ox}WL$, it should be remembered that effect of the gate voltage on the concentration of carriers will not be the same everywhere in the channel. We have seen in eqn. (4.1) that the mobile charge *per unit area* at some point along the channel is given by $C_{ox}(V_{GS} - V - V_T)$, where V is the voltage at that point (relative to the bulk or substrate). This expression can be integrated with respect to area to give the charge Q_C in the whole channel*. Then, differentiating the resulting expression for Q_C with respect to V_{GS} gives the result:

$$C_{gs} = \frac{\partial Q_C}{\partial V_{GS}} = \frac{2}{3} C_{ox} WL \qquad \text{(in saturation)} \tag{4.9}$$

*See earlier footnote regarding Q_C on page 136. For details of the calculation see e.g. p. 244–6 of the book by Pulfrey & Tarr mentioned in the list of further reading.

Between the linear and saturation regions of operation, both C_{gs} and C_{gd} vary smoothly in a predictable way, as sketched in Fig. 4.14. In the figure, the constant overlap capacitances have been included in both C_{gs} and C_{gd} for completeness.

As in the unsaturated region covered in (a) above, the gate-substrate capacitance C_{gb} is negligible.

(c) *Below threshold*, when $V_{GS} < V_T$, no mobile charge exists in the channel and the transistor is *off*. However, a charge must be present on the gate to balance the charge Q_{DEP} in the depleted region below the semiconductor surface. The current which supplies this charge flows into an equivalent circuit component modelled by a capacitance C_{gb}, between the gate and the substrate connection. It consists effectively of the oxide capacitance, in series with the capacitance of the depletion layer below the Si–SiO_2 interface (see Fig. 4.13(b)), so that its value decreases as the depleted region widens. Since for $V_{GS} < 0$ there is no depletion layer, and the channel region is a *p*-type conductor, then

$$C_{gb} \to C_{ox}WL \text{ for } V_{GS} < 0 .$$

For positive values of V_{GS}, C_{gb} falls with rising V_{GS} until at $V_{GS} = V_T$ where no further changes in the depletion layer occur, and C_{gb} becomes negligible, effectively zero.

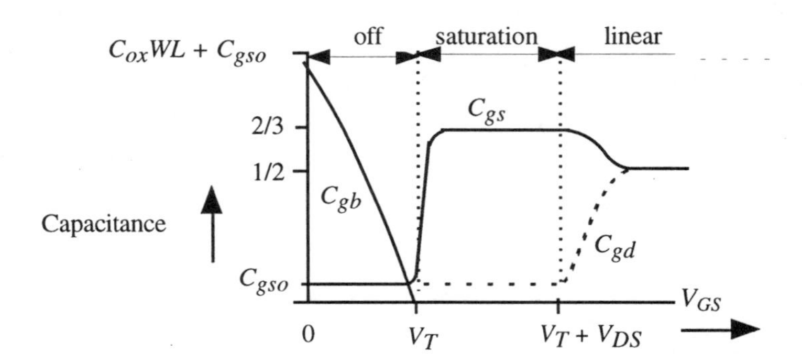

Fig. 4.14 Variation of the three gate capacitances of the Meyer model when the gate-source voltage is changed, while V_{DS} is held constant. C_{gdo} and C_{gso} are assumed to be equal.

The capacitances C_{gs} and C_{gd} have their overlap values only, since the channel is totally devoid of charge when $V_{GS} < V_T$.

Other capacitances which have not been mentioned above may affect the circuit behaviour, depending on how the transistor is connected into a circuit. These include the junction capacitance between the source and the substrate, and between the drain and the substrate. These may have values comparable to the above capacitances, but they only affect behaviour if the voltages across them change during circuit operation.

4.9 Computer models for MOSFETs

To model the electrical characteristics of a MOSFET in a computer program such as SPICE, eqns. (4.4) and (4.5) for the drain current below saturation and in saturation, respectively, are commonly used, with various modifications to allow for second-order effects that have mostly been ignored in the above treatment. Some are mentioned below.

The transistors used in very large scale integrated circuits (VLSIC) have such short channel lengths that several of the approximations made in this chapter are no longer justified. Computer models have become more sophisticated to allow for this. Nevertheless, many models (including SPICE) use most of the parameters listed below, which have been introduced in this chapter, to characterise the behaviour of each transistor. The user of the program is usually free to choose the values of these parameters, or to have them calculated from more basic data.

C_{ox} – the gate oxide capacitance per unit area.
μ_e, μ_h – the mobility of the carriers in the channel when the gate voltage has the threshold value.
W, L – the channel width and length.
V_T – the threshold voltage.
V_{DEP} – the voltage across the depletion layer at threshold, used in eqn. (4.11).
C_{gso}, C_{gdo} – the capacitances due to the 'overlap' of the gate with the source and the drain, respectively.
C_j – the junction or depletion-layer capacitance per unit area between the source and substrate, or between the drain and substrate.

N_{sub} – the doping concentration in the substrate, usually N_A in this chapter.
V_A – the Early voltage, characterising channel length modulation.
g_{fs} – the small-signal transconductance.
C_{gs}, C_{gd}, C_{gb} – the interelectrode capacitances shown in Fig. 4.12.

4.10 Magnitude of the threshold voltage

Up to this point we have taken the threshold voltage V_T for granted, and it is now time to look into what controls its value. To find the various contributions to the gate voltage at threshold, let us reconsider the energy-band diagram at threshold, which was first drawn in Fig. 4.4(b) and is shown again in Fig. 4.15. We shall denote the charge on the gate at threshold as Q_{GT}, and as usual, the mobile charge in the channel is assumed to be zero.

The gate-source voltage at threshold is equal to the gate-substrate voltage, since the source and substrate are assumed to be connected. This voltage is represented in the figure by the energy difference between the Fermi levels in the gate and substrate regions. The voltage can be seen to be the sum of two terms:

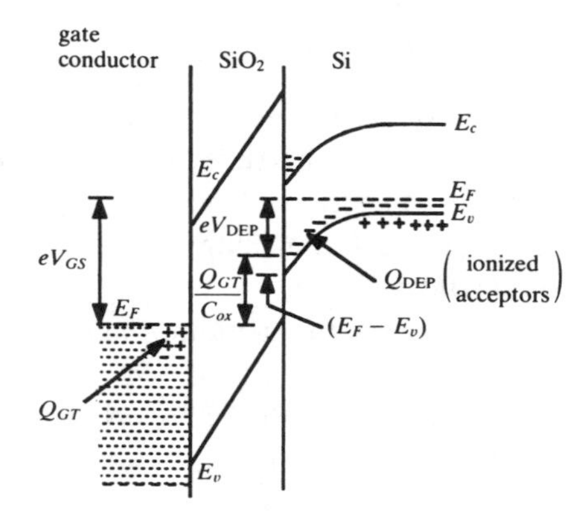

Fig. 4.15 Energy band diagram drawn at threshold, showing the two principal contributions V_{DEP} and Q_{GT}/C_{ox} to the potential difference between the gate and the substrate.

(i) a voltage across the oxide, which is just given by Q_{GT}/C_{ox}, and
(ii) a potential difference V_{DEP} across the depletion layer lying below the channel.

Thus we have as a fundamental starting-point:

$$V_T = V_{DEP} + \frac{Q_{GT}}{C_{ox}}$$

In the simplest case, the gate charge at threshold Q_{GT} just balances the charge in the depletion layer which we have called Q_{DEP} in earlier sections, as there are no other charges present to consider. Then in the equation above, we can put $Q_{GT}/C_{ox} = - Q_{DEP}/C_{ox}$, giving

$$V_T = V_{DEP} - \frac{Q_{DEP}}{C_{ox}} \tag{4.10}$$

Now in any uniformly doped *p*-type depletion layer, the charge per unit area is related simply to the charge it contains per unit volume, $-eN_A$, and its thickness W_p. Thus, $Q_{DEP} = - eN_AW_p$, so that eqn. (4.10) becomes

$$V_T = V_{DEP} + \frac{eN_AW_p}{C_{ox}}$$

The depletion layer thickness W_p was related to the potential difference across it in section 3.7. Using eqn. (3.14) and remembering that V_{DEP}, as used here, is simply the potential difference across the depletion layer, we find the relation:

$$W_p = \sqrt{\left(\frac{2\varepsilon V_{DEP}}{eN_A}\right)}$$

Here, it should be noted that ε is the permittivity of the silicon in the depletion layer, and *not* that of the oxide from which the gate insulator is made. Inserting the expression for W_p into the equation for V_T gives the result we have been seeking:

$$V_T = V_{DEP} + \frac{\sqrt{2\varepsilon eN_AV_{DEP}}}{C_{ox}} \tag{4.11}$$

To estimate V_{DEP}, note that its value is just sufficient to bring the Fermi level E_F at the semiconductor surface close enough to the edge E_c of the conduction band for the surface to become *n*-type, just as the built-in potential across a *p–n* junction is sufficient to ensure that E_F is close to E_c on one side of the junction and just as close to E_v on the other. In other words, the value of V_{DEP} is just about equal to the built-in voltage V_o of a *p–n* junction with the *same doping concentration*. It is thus about 0.7–0.8 V, as found in Chapter 3, and may be calculated using eqn. (3.2).*

This enables us to estimate V_T using eqn. (4.11). Taking the *p*-type substrate doping concentration to be $N_A = 10^{22}\ \text{m}^{-3}$, we find that $V_{DEP} = 0.7$ V, $W_p = 0.3\ \mu\text{m}$, and $Q_{DEP} = -5 \times 10^{-4}\ \text{C/m}^2$. Hence, since $C_{ox} = 7.0 \times 10^{-4}\ \text{F/m}^2$ when the oxide thickness is 50 nm, we estimate that

$$-\frac{Q_{DEP}}{C_{ox}} = \frac{\sqrt{2\varepsilon e N_A V_{DEP}}}{C_{ox}} = 0.7\ \text{V}$$

Thus we find that $V_T = 0.7 + 0.7$ V + any neglected contributions. In other words, V_T is about 1.4 V.

In practice, V_T is adjusted during fabrication by controlling V_{DEP}, which is sensitive to the doping concentration at the surface of the semiconductor. We have already noted that V_{DEP} is the amount that the energy bands have to be shifted in Fig. 4.15 in order to obtain inversion of the semiconductor type at the surface. If we inject a very thin layer of donor atoms into the surface to make it less *p*-type, then at zero gate voltage, the conduction band edge E_c at the surface would lie nearer to the Fermi level, as in Fig. 4.16. This means that a smaller applied voltage is needed on the gate to make the surface *n*-type. The change in V_T produced by an infinitesimally thin layer of donors at a surface concentration of N_S per unit area is readily shown to be equal to eN_S/C_{ox}. The extra dopant must only be introduced in a thin layer beneath the gate oxide and not elsewhere. Otherwise the drain and source regions would not be well insulated from one another, or from other transistors on the same wafer. The usual means of doing this is termed ion implantation, and is described in chapter 6.

*To define V_{DEP} at threshold, the electron concentration in the channel is normally taken to be equal to the hole concentration in the bulk. Thus eqn. (3.2) leads to $V_{DEP} = \frac{kT}{e} \ln \frac{N_A^2}{n_i^2}$

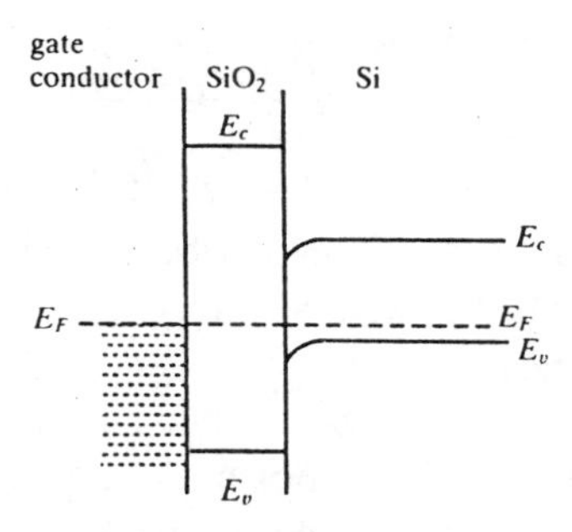

Fig. 4.16 Energy band diagram showing an unbiased *n*-channel transistor in which the semiconductor surface is lightly doped *p*-type to reduce the threshold voltage.

Before concluding, it is worth noting that those contributions which have been neglected in the derivation of eqn. (4.11) for V_T include the following:

(i) Positive charge fixed at the oxide-silicon interface, for example, ions of sodium or other metals accidentally included in the glassy SiO_2. The effect of these is small in a modern fabrication process, and amounts to a correction of about -0.1 V in an *n*-channel FET.

(ii) A built-in potential difference between the gate and the semiconductor may arise if the materials used for each are different. This is a small contribution (~ 0.1 V) in those cases where the gate is made of effectively metallic "polysilicon"* of the same doping type (i.e. *n*- or *p*-) as the substrate. But most commonly, the gate material which is chosen for use in *n*-channel transistors (made on a *p*-type substrate) is *n*-type polysilicon. Then a correction to V_T of about -0.8 V may be needed, i.e. V_T is lowered by this amount (in a *p*-channel transistor using an *n*-type polysilicon gate, the threshold is *raised* by about 0.1 V).

(iii) A potential difference between the substrate and the source also affects V_T, as explained in Panel 4.3.

---

*See Chapter 6.

PANEL 4.3

The Body Effect in MOSFETs

When the semiconductor substrate (or bulk, or body) of the transistor is not short-circuited to the source, but is allowed to take a different potential, V_{BS}, the threshold voltage of the transistor is altered. This is known as the BODY EFFECT. Since the source-substrate *n–p* junction must not be forward biased in a normal operation, V_{BS} may have one sign only (negative in an *n*-channel transistor), and the magnitude of the threshold voltage is invariably *raised* relative to its value when $V_{BS} = 0$.

The action of the substrate is sometimes likened to that of a second gate electrode (a "back-gate") which tends to turn off the transistor as the substrate voltage relative to the source is raised in absolute magnitude. However, the change in threshold voltage is not linearly related to V_{BS}, since it affects only the charge per unit area in the depletion layer beneath the channel and nothing else. Hence eqn. (4.11) may be modified appropriately to give the correct threshold voltage by replacing V_{DEP} by $(V_{DEP} + V_{SB})$ just under the square root, giving the result

$$V_T = V_{DEP} + \frac{\sqrt{2\varepsilon e N_A (V_{DEP} + V_{SB})}}{C_{ox}}$$

In the small-signal equivalent circuit, a small-signal voltage v_{sb} causes a small-signal drain current which is approximately proportional to v_{sb}, so that the action of the substrate may be represented by an additional controlled current generator of magnitude $g_{fb}v_{sb}$, in parallel with the existing generator $g_{fs}v_{gs}$ shown in Fig. 4.10.

4.11 Other types of silicon MOSFET

The p-channel MOSFET

It is easy to see that a *p*-channel MOSFET can be made in a similar way to the *n*-channel device described in this chapter. It is merely necessary to interchange *p*-type for *n*-type dopants, and vice-versa. In order to attract holes to the surface of the semiconductor and create a *p*-channel at the surface of the *n*-type substrate, a *negative* potential is needed on the gate. Thus the threshold voltage is negative in this enhancement transistor. In a normal operation, holes travel from the source to the drain, which must therefore be negatively biased relative to the source (the drain is by definition the more *negative* terminal in this case). Since, by convention, a current flowing *into* the drain has a positive sign, the drain current in the *p*-channel device is *negative*. All currents and voltages are thus reversed, as can be seen from the characteristics of a typical *p*-channel enhancement transistor, shown in Fig. 4.17(b). The input characteristic — that is, a plot of I_D versus V_{GS} when V_{DS} is held constant — and the circuit symbol are also shown. For comparison, Fig. 4.17(a) has the *n*-channel equivalents.

Equations (4.4) and (4.5) give the correct magnitude (but the wrong sign) for I_D in the *p*-channel MOSFET, without making any changes, but remembering nevertheless that normally all of the voltages must be given negative values. Since the mobility of holes is about half that of electrons, the conductance of the *p*-channel is correspondingly lower, and so is the transconductance g_{fs}, as can be seen from eqn. (4.6) (sometimes the ratio W/L is raised by the designer to compensate). Other small-signal circuit components can be derived using the same equations as given above. The above expressions for the threshold voltage can only be used after appropriate changes of sign.

Depletion-mode MOSFETs

Since the threshold voltage can be lowered by implanting dopants in the channel region, it is possible to reduce V_T to zero, and even to reverse its sign. An *n*-channel MOSFET with a negative threshold voltage will have a conducting channel for all gate voltages which are more positive than V_T. So it will be conducting at zero gate bias, and negative gate voltages will cause the channel to become *depleted* of electrons. This device is called a DEPLETION-MODE FET, and its symbol and characteristics are shown in Fig. 4.17(c). Since only V_T is affected by implanting dopants into the channel, the input characteristic, I_D vs V_{GS}, is simply a copy of the enhancement transistor's characteristic, but shifted along the V_{GS} axis.

A *p*-channel depletion-mode transistor can be made in a similar way, and its characteristics are shown in Fig. 4.17(d).

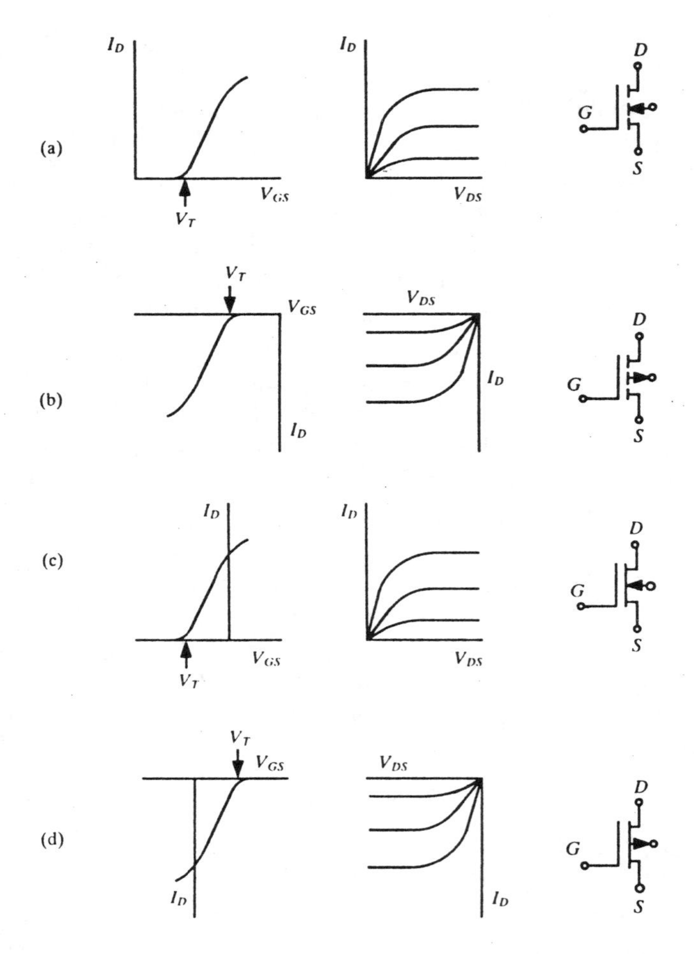

Fig. 4.17 Gate transfer characteristics (left), output characteristics (centre) and international symbol (right) for each of the four principal kinds of transistor: (a) *n*-channel enhancement (b) *p*-channel enhancement (c) *n*-channel depletion (d) *p*-channel depletion transistor.

Summary of Terminology for MOSFETs

Threshold voltage – the minimum gate voltage for formation of a conducting channel.

Channel – the region of high conductivity between source and drain.

Oxide capacitance	–	capacitance per unit area of the gate-oxide-conducting-channel sandwich, when $V_{DS} = 0$.
Transconductance	–	small-signal short-circuit output current for unit small-signal input voltage.
Common-source connection	–	the source terminal is common to input and output voltages.
Unsaturated, or Triode region	–	region of output characteristics where $\lvert V_{DS}\rvert < \lvert V_{GS} - V_T\rvert$.
Saturation region	–	region of output characteristics where $\lvert V_{DS}\rvert \geq \lvert V_{GS} - V_T\rvert$.
Transit time	–	the time required for a carrier to travel the length of the channel.
Body effect	–	the change in threshold voltage resulting from a change in source-substrate voltage.
Pinch-off voltage	–	the drain-source voltage which results in depletion of carriers in a region in the channel.
Early voltage	–	the increase in V_{DS} which results in a 100% increase in the saturation drain current.

PROBLEMS

4.1 In an *n*-channel MOSFET, is the channel conducting when the gate voltage is (a) greater than threshold (b) less than threshold (c) below saturation (d) in saturation?

4.2 Saturation in an *n*-channel MOSFET occurs under which conditions: (a) $V_{GS} > V_{DS}$ (b) $V_{GS} < V_{DS} - V_T$ (c) $V_{GS} - V_T > V_{DS}$ (d) $V_{DS} > V_{GS} - V_T$ (e) $V_{DS} > V_{GS}$?

4.3 How do the answers to 4.1 and 4.2 differ in the case of a *p*-channel enhancement MOSFET?

4.4 In what conditions is the channel conductance of an FET dependent only on V_{GS}?

4.5 In the saturation region, does the channel become (a) depleted of carriers near the drain (b) depleted of carriers near the source (c) non-conducting near the drain (d) non-conducting near the source?

4.6 Draw the small-signal equivalent circuit of a MOSFET and explain the physical origins of each component.

4.7 List all the factors which play a part in determining the threshold voltage of a MOSFET. Explain how its value can be adjusted during manufacture.

4.8 Draw a labelled diagram showing the formation of an inversion layer in a silicon *p*-channel enhancement MOSFET having an n-type polysilicon gate, when V_{DS} is zero.

4.9 Explain why the electron mobility in the drain current eqn. (4.4) is about one half of the mobility in bulk silicon.
The measured transconductance g_{fs} of such a FET at a saturation drain current of 1.4 mA is 0.5 mS. The channel length and width are 5 μm and 20 μm respectively. Estimate the value of the small-signal input capacitance of the transistor in saturation.

4.10 Sketch the output characteristics of a *p*-channel enhancement MOSFET. Deduce and explain their shape in terms of the charges induced by the gate in the channel. Explain how it can be turned into a depletion mode device by a single additional processing step.

4.11 By considering the distribution of potential and charge in the channel of a FET, explain the physical origin of the saturation of I_D. Show that in the triode region the slope of the output characteristic just reaches zero at the onset of saturation.

4.12 A silicon *n*-channel MOSFET has source-drain separation of 15 μm, gate width 50 μm and oxide thickness 100 nm. The threshold voltage is measured to be 1.5 V and the effective electron mobility is 0.08 m^2/Vs. Calculate the channel conductance at $V_{GS} = 5$ V and $V_{DS} \ll V_{GS}$. Take the permittivity of SiO_2 to be 4.0.

4.13 Draw the small-signal equivalent circuit for the FET of Problem 4.12, and estimate the values of C_{GS} and g_{fs} when operated in the saturation region at a quiescent drain current of 1 mA.

4.14 The saturation drain current for a commercial *n*-MOSFET is 2.5 mA at V_{GS} = 10 V and 7.5 mA at V_{GS} = 15 V, Estimate the threshold voltage and the value of $C_{ox}W/L$.

4.15 Figure P.4.1 shows a digital inverter circuit, in which it is required that the output voltage should be zero when the gate voltages are zero. If the threshold voltages are 2.0 V and 2.5 V for the *n*- and *p*-channel devices respectively, and $\mu_e = 2\mu_h$, find the ratio of the gate widths of the two devices, the channel length and oxide thicknesses being equal.

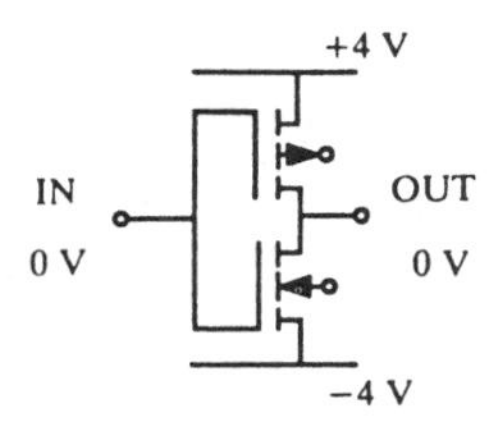

Fig. P.4.1

4.16 Deduce and sketch the $I_D - V_{DS}$ characteristic for the MOSFET of problem 4.14 when the gate is connected to the drain.

4.17 Starting with eqn. (4.3), show that the potential V at a point in the channel varies parabolically with x, and sketch the graph of V against x. How does the carrier velocity vary with x?

4.18 Why must the gate of a MOSFET overlap both the source and drain diffusions slightly, and what effect has this overlap on the equivalent circuit?

4.19 Calculate the transit time in an *n*-channel FET having channel length 1.4 μm and width 5 μm, assuming a saturation velocity equal to the value in intrinsic Si, namely 10^5 m/s. Hence, show that the input time constant limits the switching speed when the oxide thickness is 40 nm, the driving voltage has a source resistance of 1100 Ω, and the capacitance between the gate input conductor and substrate is 0.03 pF.
Does the same apply in the absence of the input conductor capacitance?

4.20 With the aid of Fig. 4.6, explain why the capacitance per unit area between the gate and the source varies along the channel.

4.21 Explain why is it impossible to reduce overlap capacitances to negligible values. Measurements on a symmetrical *p*-channel MOSFET give the following results: Gate width $W = 2.5$ μm, gate length $L = 0.8$ μm, and, when $V_{GS} = V_{DS}$, the capacitances $C_{gs} = 1.6 \times 10^{-3}$ pF and $C_{gd} = 2 \times 10^{-4}$ pF. Estimate the oxide thickness and the value of the transconductance in saturation at a drain current of 0.4 mA. You may assume $\mu_h = 0.03$ m^2/Vs

4.22 The circuit diagram, including substrate connections, of a digital inverter is shown in Fig. P4.1. Redraw it, including on it all capacitances associated with the *p–n* junctions used to isolate the sources and drains from the substrates. Which of these is likely to affect the speed at which the inverter operates?

4.23 The threshold voltage may be expressed as the sum of four contributions: which are the two most important for p-channel MOSFETs having n^+ doped polysilicon gates? An enhancement transistor of this type in a wafer doped with 1.5×10^{22} acceptors/m^2 is intended to have a threshold voltage of -0.9 V. Estimate

(i) the oxide capacitance required, and hence its thickness,

(ii) the implanted dopant density required in the channel to adjust the threshold voltage to -1.2 V.

Chapter 5

The Bipolar Transistor

5.1 Construction

In the bipolar transistor, two parallel *p–n* junctions are formed as close together as possible to make either an *npn* or a *pnp* transistor. Using the planar diffusion process as described in Chapter 6, the two structures shown in Fig. 5.1 are commonly made. The uppermost region is called the EMITTER, the lowest is the COLLECTOR, and the two are separated by the BASE, which is made as thin as possible (~1 μm), so that minority carriers flowing across it from emitter to collector have very little chance of recombining on the way with carriers of the opposite sign.

Although the doping profiles in diffused transistors are not abrupt, we shall assume that the change in doping across each junction occurs abruptly, as in Chapter 3, in order to deduce straightforward expressions for the currents.

Throughout this chapter we shall concentrate upon the *pnp* transistor, for convenience — there are no different principles involved in the operation of an *npn* device although the signs of all voltages and currents are reversed, and of course electrons are replaced by holes, and vice versa.

A thorough understanding of the *p–n* junction diode is advised before tackling this chapter.

5.2 Principles of operation

When used as a current amplifier, the two junctions forming the transistor are differently biased. The emitter-base junction is in *forward* bias, whilst the collector-base junction is *reverse* biased.

Consider first the current flowing across the emitter-base junction, as if it were simply a forward-biased diode. The *p*-side (the emitter) is made much more heavily doped than the base, so that the forward current across the junction is carried predominantly by holes flowing into the base as illustrated in Fig. 5.2.

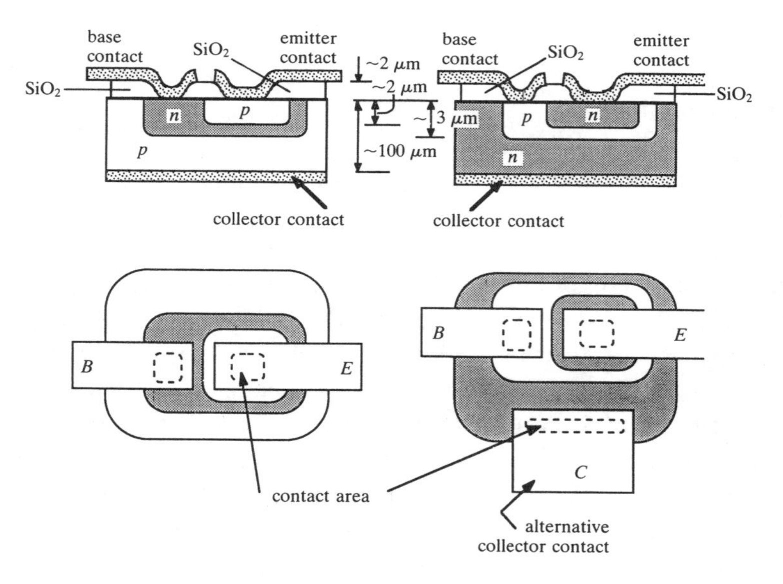

Fig. 5.1 Sectional (top) and plan views of *pnp* and *npn* transistors made by diffusion of impurities from the upper surface of a wafer. The bulk of the wafer forms the collector region. The base is made next by diffusing impurities of the opposite type into a rectangular region. Then the emitter is made by diffusing appropriate impurities into a smaller rectangle within the base, to a shallower depth. Contacts are made through windows etched in SiO_2 insulation, by depositing a thin layer of a metal such as aluminum.

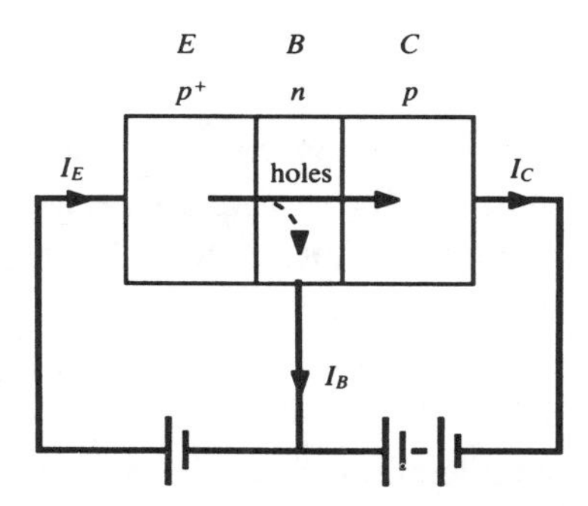

Fig. 5.2 Hole current flow and bias voltages in a *pnp* transistor.

Now we know from the theory of the diode in Chapter 3 that if a contact is placed on the base side close enough to the junction, few holes recombine before they reach the contact. In place of a metal contact we use the *p*-type collector region, biased negatively with respect to the base in order to encourage holes to cross to it from the base (Fig. 5.2). Because of the reverse bias across this junction there is a strong electric field in it, which sweeps away all holes which diffuse to the edge of the depletion layer from the base.

The difference between the collector-base junction and a reverse-biased diode is that the latter has a negligible supply of holes in the *n*-type region, where they are minority carriers. In the transistor the hole concentration is artificially increased by injection from the emitter. The current across the collector-base junction can therefore be much larger than the reverse saturation current or leakage current of the junction. Indeed, since nearly all of the holes which leave the emitter cross into the collector, the collector current I_C equals to a first approximation the current I_E flowing in the emitter-base junction. Thus, as we shall see,

$$|I_C| \cong I_E \propto \exp\frac{eV_{EB}}{kT}$$

However, this description neglects the small but important current I_B flowing out of the base contact, which we shall show is directly proportional to the collector current I_C.

5.3 Currents in the base

Current conservation (Kirchhoff's current law) tells us that the current I_B flowing *out* of the base must be the difference between the current flowing into the emitter, and that flowing out of the collector, i.e.

$$|I_E| = |I_C| + |I_B|$$

(In studying the internal operation, it is easier to avoid the sign conventions used in circuit theory, which require that *inward* currents have positive signs.) There are two main contributions to I_B, for which equations must be found, and one minor contribution to be discussed:

(i) Part of the emitter current consists of electrons injected *into* the emitter, which come from the base and not the collector.* We call this current I'_B in this book.

(ii) Some holes in the base recombine with electrons and do not reach the collector. Electrons flow in from the base contact to supply the recombination process, as illustrated in Fig. 5.3. We call this current I''_B.

(iii) Reverse bias across the base-collector junction causes the normal reverse saturation and leakage currents to flow between collector and base. This current, I_{CBO}, is negligible (except in germanium transistors) when the transistor is biased normally.
All of these currents are illustrated in Fig. 5.3.

Note carefully the first two base current contributions described above, which flow in the emitter, but not in the collector, while the third flows in the collector circuit but not in the emitter. Thus

$$|I_B| = I'_B + I''_B - I_{CBO}$$

We shall now find expressions for the collector current and the base current, to show that they are proportional to one another.

*The fraction of I_E injected into the base is usually called the EMITTER EFFICIENCY. Thus I'_B arises from what might be termed the inefficiency of the emitter.

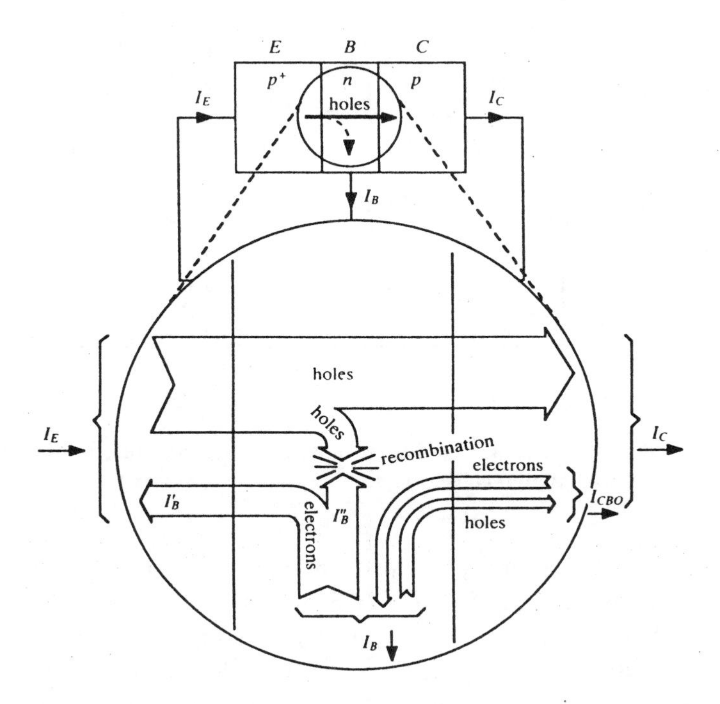

Fig. 5.3 A more detailed look at the flows of electrons and holes in the junction regions of a *pnp* transistor.

5.4 Calculation of the collector current

The collector current can be calculated from the flow of holes in the base region. Figure 5.4 shows how the concentration of holes varies across the transistor; electrons are ignored in the figure for clarity. The same bias voltages have been assumed in drawing this diagram, as are shown in Fig. 5.2.

Figure 5.4 is an important diagram, from which many of the transistor's properties can be deduced, and it should be studied carefully. The region on which we concentration is the base, whose effective width W in Fig. 5.4 is assumed to be small compared to a diffusion length, and is around 1 μm in reality.

At the emitter end of the base, the hole concentration is raise above its equilibrium value by the factor $\exp(eV_{EB}/kT)$ — just as in a p^+n diode

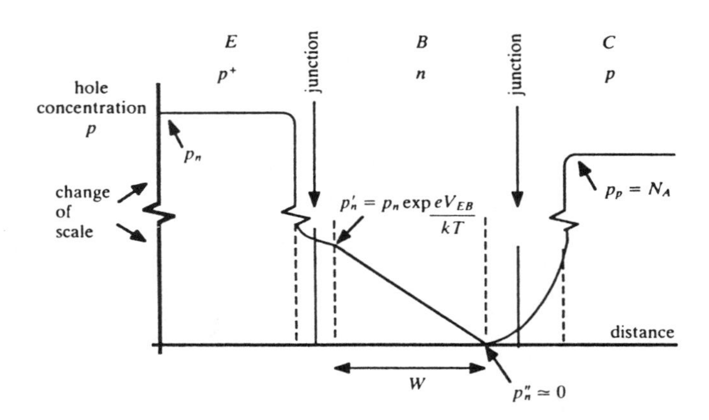

Fig. 5.4 Concentration of holes across the emitter, base and collector of a normally biased *pnp* transistor. Note the change in scale on the vertical axis.

(i.e. one which is very heavily doped on the *p*-type side). The base being thin, the hole concentration falls linearly towards the collector. Holes diffuse down this gradient, there being no drift current in this uncharged, field-free region. Since the depletion layers are normally less than 0.1 μm wide, the neutral region, of width W, extends almost across the whole width of the base.

At the collector end of the base, the hole concentration p''_n is reduced below the equilibrium value p_n by the reverse bias across the collector-base junction. The hole concentration p''_n is given by the usual expression for a reverse biased junction:

$$p''_n = p_n \exp \frac{-eV_{BC}}{kT}$$

The base-collector voltage V_{BC} is normally in excess of 0.1 V, which implies that $p''_n \ll p_n$, and we may assume that this hole concentration is zero, as shown in Fig. 5.4.

The diffusion current of holes across the base and into the collector can therefore be written using the diffusion equation:

$$I_p = -eAD_h \frac{dp}{dx} \tag{5.1}$$

where A is the junction area. As dp/dx is simply the slope of the graph in Fig. 5.4, we can rewrite eqn. (5.1):

$$I_p = eAD_h \frac{(p'_n - p''_n)}{W} = eAD_h \frac{p'_n}{W}$$

Since $p'_n = p_n \exp(eV_{EB}/kT)$ we can now write an expression for the collector current (conventionally positive *into* the transistor) ignoring I_{CBO}, which is very small

$$I_C = -I_p = -eAD_h \frac{p_n}{W} \exp\left(\frac{eV_{EB}}{kT}\right) \tag{5.2}$$

The collector current is proportional to $\exp(eV_{EB}/kT)$, making its value very sensitive to changes in the small voltage, V_{EB}. It is because of this that V_{EB} is almost never used directly to control the collector current in a circuit.

We shall now calculate the two principal base current contributions, to show that they are directly proportional to the above expression for I_C.

5.5 Base current injected into the emitter: I'_B

That part of the base current which is due to a flow of electrons from the base into the emitter, is proportional to the hole current flowing the other way, just like the electron and hole currents in a *p–n* diode. Since the emitter is usually *very* heavily doped, most electrons recombine before reaching the contact, i.e. the diffusion length $(D_e \tau_e)^{1/2}$, which will be abbreviated as L_e, is small.

In section 3.10 this case was treated (as the 'thick' diode) and it was shown that the concentration gradient dn/dx at the depletion layer edge is just $(n'_p - n_p)/L_e$, which in normal foward bias conditions is almost exactly equal to n'_p/L_e. So the diffusion current, which equals I'_B is just

$$|I'_B| = eAD_e \frac{dn}{dx} = eAD_e \frac{n'_p}{L_e}$$

(compare this with eqn. (3.26)).

Putting $n_p' = n_p \exp(eV_{EB}/kT)$ leads to the result

$$|I_B'| = eAD_e \frac{n_p}{L_e} \exp\left(\frac{eV_{EB}}{kT}\right)$$

You can see that this is proportional to I_C, by dividing it into eqn. (5.2), so finding the ratio:

$$\left|\frac{I_C}{I_B'}\right| = \frac{D_h}{D_e} \cdot \frac{L_e}{W} \cdot \frac{p_n}{n_p}$$

Provided that the doping levels $N_D^{(B)}$ in the base and $N_A^{(E)}$ in the emitter are not too high, the usual expressions $p_n = n_i^2/N_D$ and $n_p = n_i^2/N_A$ can be inserted, leading to the result

$$\left|\frac{I_C}{I_B'}\right| = \frac{\mu_h}{\mu_e} \cdot \frac{L_e}{W} \cdot \frac{N_A^{(E)}}{N_D^{(B)}} \tag{5.3}$$

where the fact that $D_h/\mu_h = D_e/\mu_e$ has also been used.

It was noted above that, if the depletion layers in Fig. 5.4 are not very wide, the width W in eqn. (5.3) is approximately equal to the width W_B of the base region. In this case, eqn. (5.3) shows that I_C/I_B' is a constant, i.e. I_B' is proportional to I_C.

Later on it will be necessary to take account of the fact that the width of the depletion layer at the collector end of the base may not always be negligible.

5.6 Base recombination current: I_B''

To calculate the recombination current, the results of section 3.9 can be used. It was shown there that an excess concentration $(p - p_n)$ of holes in an n-type region (the base) recombines at a rate $(p - p_n)/\tau_h$, where τ_h is the hole lifetime. The rate of recombination of charge in any small volume element dV is thus equal to the excess charge $dq = e(p - p_n)dV$ in that volume, divided by the lifetime τ_h. Integrating this rate over the whole base gives the recombination current I_B'':

$$I_B'' = \int_{\text{base}} \frac{dq}{\tau_h} = \frac{1}{\tau_h}\int dq, \text{ assuming } \tau_h \text{ is constant.}$$

But $\int dq$ is just the total excess minority charge in the base, which will be written simply as Q_B. Hence

$$I_B'' = \frac{Q_B}{\tau_h} \tag{5.4}$$

The minority carrier distribution shown in Fig. 5.4 can be used to calculate Q_B. For a transistor with unit cross-sectional area, Q_B is just the electron charge times the area under the graph in the base region. Neglecting p_n compared with p_n', the area concerned is just a triangle of height p_n' and base W. Hence

$$Q_B = Ae \cdot \frac{1}{2} p_n' W \tag{5.4a}$$

where A is the cross-sectional area of the transistor. Using eqn. (5.4)

$$|I_B''| = \frac{AeWp_n}{2\tau_p} \exp\left(\frac{eV_{BE}}{kT}\right) \tag{5.5}$$

Once again the ratio of the collector current and I_B'' can be formed:

$$\left|\frac{I_C}{I_B''}\right| = \frac{2D_p\tau_p}{W^2} \tag{5.6}$$

which, as before, is constant since $W \cong W_B$.

The simple way of appreciating this relationship is by noting that whilst I_B'' is proportional to the area under the graph in Fig. 5.4, I_C is proportional to its slope.

5.7 Collector-base 'leakage' current: I_{CBO}

The symbol I_{CBO} stands for the *C*ollector-*B*ase current when the third terminal (the emitter) is *O*pen circuit, and the collector is biased normally. It therefore is identical to the reverse current of an equivalent diode, and consists of electron and hole flow as indicated in Fig. 5.3. In a silicon transistor this current can usually be ignored in comparison to the other contributions to I_B, when the transistor is biased, as in Fig. 5.2, in the active region. In a germanium transistor I_{CBO} is not entirely negligible.

5.8 Current gain h_{FE}

The above results confirm that I_B is proportional to I_C, given some assumptions which boil down to supposing that carrier injection into the base is neither very small, nor very large, and that a reverse bias is maintained across the collector-base junction. The large-signal CURRENT GAIN, which is defined by the equation

$$h_{FE} = I_C / I_B$$

should thus be approximately constant, and can in principle be calculated from eqns. (5.3) and (5.6). In practice, a more complicated analysis is needed, because of the gradual doping profiles and the fact that at the surface the emitter is doped beyond the point where it behaves like a semiconductor, and is quasi-metallic. However, these equations are useful in indicating the salient factors controlling h_{FE}.

First, the base width must be small, to reduce recombination (see eqn. (5.6)). Base widths down to as little as 50 nm have been used, though more usually they are in the range 0.2–1 μm. This, coupled with the effects of the doping gradients in the base,* make the recombination current much smaller in practice than that injected into the emitter in nearly all transistors constructed as in Fig. 5.1. Thus eqn. (5.3) can be used as a rough guide for

*The doping gradient creates a built-in electric field, like that in a junction, which speeds the carriers across the base, reducing their chances of recombining.

estimating h_{FE}. Note that the high emitter doping N_A may cause the diffusion length L_e in the emitter to be less than the base width W_B. So to obtain a large current gain h_{FE}, eqn. (5.3) shows that the emitter doping N_A must be made more than a hundred times the base doping N_D. Since in a diffused transistor the emitter doping rises with distance from the junction (see Fig. 5.10), the surface doping is in practice several orders of magnitude above the highest base doping.

5.9 Transistor ouput characteristics

In discussing now the shape of a transistor's characteristics we shall introduce two features we have hitherto ignored. They are (i) that the effective base width W varies with V_{CB} owing to changes in the collector depletion layer width and (ii) the effects on I_C and h_{FE} of a forward bias across the base-collector junction.

Common base characteristics

We consider first the output characteristic, I_C vs V_{CB}, for the common-base connection (Fig. 5.5(a)), so-called because the base terminal is common to both the input terminals on the left (the emitter and base) and the output terminals on the right (the collector and base). Note that, by convention, the collector current characteristic is plotted while the input current I_E is held constant and V_{CB} is varied. As remarked above, the depletion region at the collector end of the base widens as $|V_{CB}|$ is increased, causing the effective base width W, defined in Fig. 5.4, to decrease. The distribution of holes in the base region shown Fig. 5.4 is redrawn in Fig. 5.6 to illustrate this for two different values of V_{CB}. The effect is called BASE WIDTH MODULATION. As V_{CB} varies, the graph of the hole concentration $p(x)$ moves parallel to itself from ① to ② in the figure, since $I_E \propto dp/dx = \text{constant}$. Thus p_n' must decrease, and, with it, V_{EB} and I_B — both the recombination current I_B'' and the current of electrons injected into the emitter I_B' decrease. I_C remains almost unchanged; the slight increase with $|V_{CB}|$ as I_B falls is barely noticeable. The corresponding

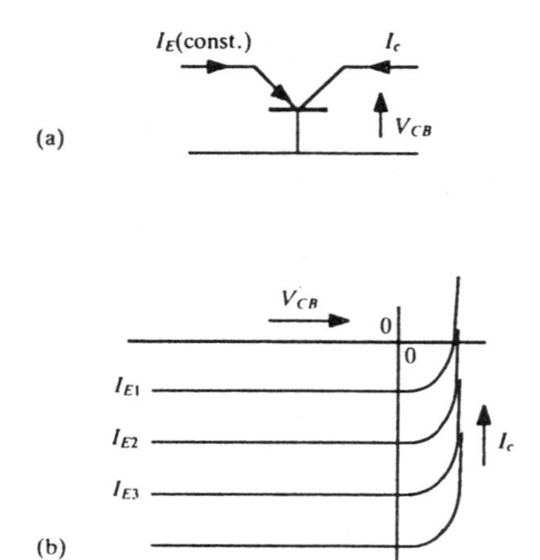

Fig. 5.5 (a) The common base connection and (b) output characteristics of a *pnp* transistor.

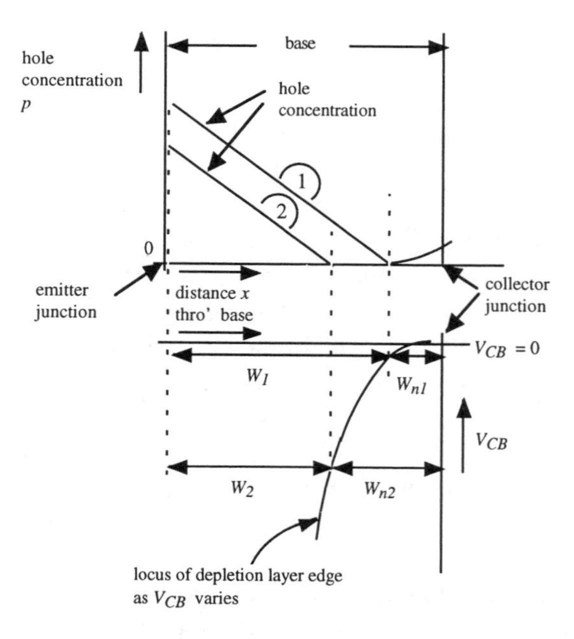

Fig. 5.6 Hole concentration distribution in the base (top) for two values of V_{CB}. The graph below shows the position of the depletion layer edge as a function of V_{CB}.

I_C–V_{CB} characteristics for various I_E values are shown in Fig. 5.5(b) to the left of the I_C axis. This part of the characteristic is said to be the ACTIVE REGION, defined by the combination of forward bias on the base-emitter junction with reverse bias on the collector.

Now consider zero and positive values of V_{CB}, the collector junction bias.

When $V_{CB} = 0$, the concentration p_n'' in Fig. 5.4, instead of being zero, is equal to the equilibrium value p_n; again I_C is little changed, as shown on the characteristic in Fig. 5.5(b). But if V_{CB} is made *forward* biased (i.e. $V_{CB} > 0$), p_n'' rises substantially as in Fig. 5.7, and may become greater than was p_n' beforehand. The base current I_B must now be greatly increased — both by additional recombination in the base and by extra injection into the emitter. Once I_B becomes a significant fraction of the emitter current I_E, then I_C must fall. It falls rapidly with increasing forward collector bias, as shown on the right of the vertical axis in Fig. 5.5(b). The shape of the minority carrier distribution in Fig. 5.7 also reflects this fall in collector current I_C relative to the emitter current I_E. The current of holes into the collector is proportional to the slope *dp/dx* in Fig. 5.7 at the point where the current enters the collector, i.e. on the right of the base. (Remember

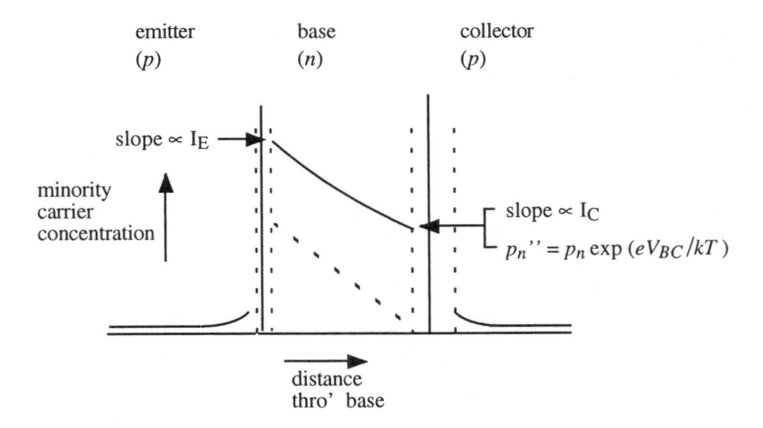

Fig. 5.7 Minority carrier concentration (holes) in the base in a saturated transistor (solid line), compared with the case for the same emitter current in the active regime (dashed line).

that the hole current in the base is purely a diffusion current.) The emitter current is likewise proportional to dp/dx at the *emitter* end of the base, i.e. on the left of Fig. 5.7, where the slope is higher.

The region of rapidly falling magnitude of the current $|I_C|$ in the characteristic curves shown in Fig. 5.5(a) is called the SATURATION region. Note that it is not a region of constant current, but of almost constant base-collector voltage.

The fall in $|I_C|$ as V_{CB} goes into forward bias can be modelled mathematically in a rather straightforward way. The collector current can be regarded as the sum of two components.

(1) The flow of holes originating from the *emitter*, through the base, and passing into the collector. This is proportional to, but slightly less than, the emitter current I_E.
(2) Holes injected in the reverse direction by the *collector*, into the base which (mostly) travel into the emitter. The collector in forward bias naturally acts like an emitter.

The principle of superposition can then be used to combine the currents, provided these two components are independent of one another. To ensure this, the first is evaluated assuming zero bias on the collector, while the second is calculated when zero emitter current flows.

The first of these currents is just the emitter current minus the base current, which flows when the transistor operates normally in the active region (h_{FE} may be assumed independent of collector bias voltage). The contribution to I_C is

$$I_E - I_B = I_E - \frac{I_E}{1+h_{FE}} = I_E\left(\frac{h_{FE}}{1+h_{FE}}\right)$$

The second current component is just given by the 'diode equation' for the collector-base junction, when the emitter is open circuit, and is

$$I_{CBO}\left(\exp\frac{eV_{BC}}{kT} - 1\right)$$

Adding these currents (remembering that they are in opposite directions) gives the following expression for the collector current

$$I_C = -I_E\left(\frac{h_{FE}}{1+h_{FE}}\right) + I_{CBO}\left(\exp\frac{eV_{BC}}{kT} - 1\right) \tag{5.7}$$

Note that this reduces to the equation which applies in the active region when the base-collector bias V_{BC} is negative and large compared to kT/e, thus:

$$I_C = -I_E\left(\frac{h_{FE}}{1+h_{FE}}\right) - I_{CBO}$$

Equation (5.7) successfully models the shape of the characteristics in Fig. 5.5(b) in both the saturation and active regions of the plot. It is widely used as a basis for computer modelling, although there is no allowance in it for the effects of base-width modulation. However, this defect can be remedied by allowing both I_{CBO} and h_{FE} to depend on V_{BC}, in a fairly simple way.

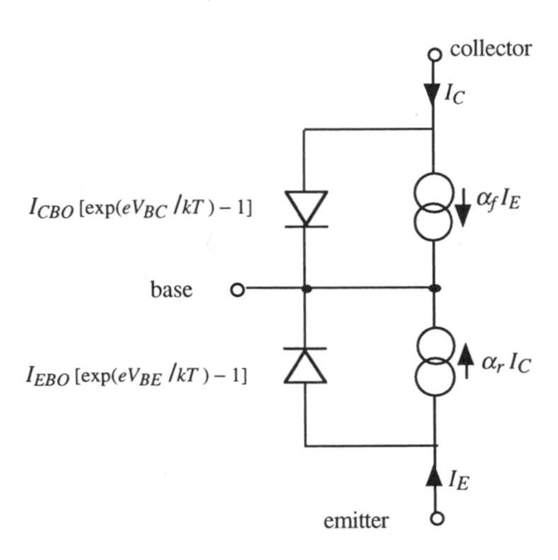

Fig. 5.8 A circuit representation of eqn. (5.7).

Equation (5.7) can also be represented as a circuit model. The collector current is made the sum of two branch currents which join at the collector terminal, as shown in Fig. 5.8. Each branch current represents one of the two terms in the equation. The ideal diode connected to the collector has the reverse saturation current I_{CBO}. Because the collector can also act as an emitter, a second controlled current generator proportional to I_C is needed to complete the model and to give the two contributions to the emitter current correctly. This full model for the bipolar transistor is usually termed the Ebers-Moll model. We shall return to it in section 5.11.

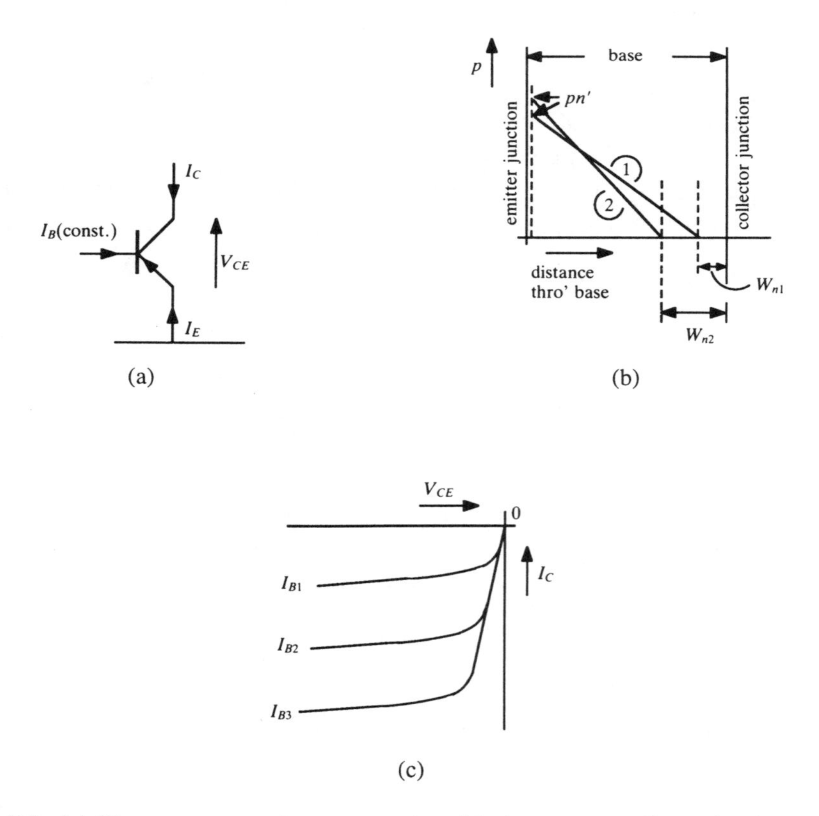

Fig. 5.9 (a) The common emitter connection (b) the corresponding minority carrier concentration across the base for two values of V_{CE} (c) the output characteristics.

Common emitter output characteristics

To obtain the common emitter characteristic we plot I_C vs V_{CE} with I_B held constant: the circuit is as in Fig. 5.9(a). As V_{CE} rises, V_{BE} varies but little, and nearly all of the increase in voltage V_{CE} appears across the collector base junction. The increase in V_{CB} causes the depletion layer in the base to widen as before, but with a different effect on the minority carrier distribution, as shown in Fig. 5.9(b). To keep I_B constant, p_n' must rise slightly to compensate for reduced recombination in the base. The slope $|dp/dx|$ rises, and so does the collector current. The output characteristics in Fig. 5.9(c) show the rise clearly.

We have already seen how $|I_C|$ drops when V_{CB} goes into *forward* bias — this occurs when $V_{CE} < V_{BE}$ in the common emitter connection, and the effect is shown clearly in Fig. 5.9(c). The reader is left to sketch the corresponding minority carrier distribution in the base.

Again, this region of rapidly falling $|I_C|$ is called the *saturation* region of the common-emitter characteristics. The collector-emitter voltage lies in the range 0.1 V–0.2 V here. It is important for the reader to realise that the term saturation is used quite differently in the bipolar transistor from its usage in the behaviour of the MOSFET, and that care must be taken to use the term correctly in each case.

5.10 The small-signal equivalent circuit

A small-signal equivalent circuit such as that shown in Fig. 5.10(a) is used to represent the behaviour of a transistor when it is employed to amplify small currents or voltages at the input terminals. We shall concentrate exclusively on this circuit, which is normally used to model the transistor operating in the active region when used in the common-emitter connection shown earlier in Fig. 5.9(a). The simplest form of the common emitter equivalent circuit is shown in Fig. 5.10(a). It models the transistor at low signal frequencies. The various components it contains are discussed next, and a simple expression for the magnitude of each is derived, by making use of the equations derived above.

To model the behaviour of a transistor at frequencies above a few kilohertz correctly, several capacitors must be added to the DC equivalent circuit.

They are shown in Fig. 5.10(b), and will be discussed after completing the treatment of the low-frequency model.

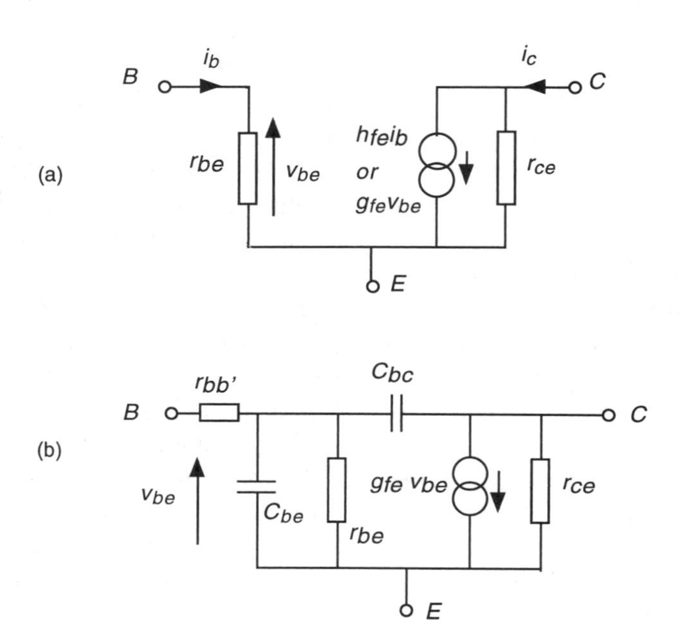

Fig. 5.10 Small-signal equivalent circuits for the bipolar transistor. (a) Low-frequency model (b) High-frequency model with capacitors added.

Current generator $g_{fe}v_{be}$ or $h_{fe}i_b$

The current generator between collector and emitter in Fig. 5.10(a) models the fact that a small change i_b in the base current I_B results in a small change i_c in the collector current. The small-signal current gain h_{fe} is defined as the ratio i_c/i_b. It can be seen from the circuit that this is only the case when no signal current flows in the resistor r_{ce}, i.e. when V_{CE} is constant. Thus:

$$h_{fe} = \frac{i_c}{i_b} = \left(\frac{\partial I_C}{\partial I_B}\right)_{V_{CE} = \text{const}} \tag{5.8}$$

h_{fe} differs from the large-signal current gain $h_{FE} = I_C/I_B$, which was discussed in section 5.8, only because real transistors do not behave precisely as described by the analysis there. Note that we use small letters and subscripts to signify the small-signal variables.

In practice, h_{fe} is not an easy parameter to control accurately during manufacture. Equations (5.3) and (5.6) show that the current gain is dependent on the base width W, which is about 1 μm or less in fast transistors. Small but significant variations in W from one microcircuit to another are possible. Because of this, it is more useful to define the strength of the current generator in Fig. 5.10(a) in terms of a small change v_{be} in the base-emitter voltage, rather than in terms of the base current i_b. The dependence of I_C on V_{BE} is strong (see eqn. (5.2)), but it is reliably and accurately predictable in real transistors.

Hence we define the *transconductance* g_{fe}[1] by the equation

$$g_{fe} = \left|\frac{i_c}{v_{be}}\right| = \left(\frac{\partial I_C}{\partial V_{EB}}\right)_{V_{CE} = \text{const}} \tag{5.9}$$

Thus g_{fe} is defined in exactly the same way as the transconductance of a MOSFET.

Note that the voltage V_{CE} is held constant in order that none of the small-signal current generated can flow in the output resistance r_{ce} shown in Fig. 5.10(a). An expression for g_{fe} can be found by differentiating eqn. (5.2) to obtain

$$g_{fe} = \frac{eI_C}{kT} \tag{5.10}$$

Notice that the transconductance is directly proportional to the collector current I_C. Hence provided I_C is well controlled by good circuit design, the current generator has a well-defined strength.

At a temperature of 290 K, eqn. (5.10) implies that $g_{fe} = 40\, I_C$, a result which the circuit designer uses as a rule of thumb (clearly the thumb must be

[1]The alternative symbol g_m in place of g_{fe} is more popular, but does not conform to internationally agreed convention.

at 290 K). The invaluable expression for g_{fe} given in eqn. (5.10) is accurate enough to be used reliably in transistor circuit design.

Input resistance r_{be}

The small-signal input resistance r_{be} relates V_{BE} and I_B through the equation

$$r_{be} = \left(\frac{\partial V_{BE}}{\partial I_B}\right)_{V_{CE}} = \left(\frac{\partial V_{BE}}{\partial I_C}\right) \cdot \left(\frac{\partial I_C}{\partial I_B}\right)$$

Thus with the help of eqns. (5.8) and (5.9) we have

$$r_{be} = \frac{h_{fe}}{g_{fe}} \tag{5.11}$$

where h_{fe} is the small-signal current gain $\frac{\partial I_C}{\partial I_B}$, evaluated with constant V_{CE}. This equation is frequently useful.

If the current gain is nearly the same for all values of collector current, then $\frac{\partial I_C}{\partial I_B} \cong \frac{I_C}{I_B}$ and as a result,

$$r_{be} \cong \frac{kT}{eI_B}$$

Thus, when $T = 290$ K, $r_{be} = \frac{0.025}{I_B}$. Unfortunately, this simple equation is not always accurately obeyed in practice because of the fact that h_{fe} varies as I_C changes in a real transistor. However, it allows us to estimate an approximate value for r_{be}. For example when $I_B = 25\ \mu A$, we estimate r_{be} to be 1 kΩ. A few kilohms is typical for the input resistance of a bipolar transistor, unless the base current I_B is particularly small.

Output resistance r_{ce} (Base width modulation, or the Early effect)

The output resistance r_{ce} in the common-emitter connection reflects the changes in external collector current I_C with the collector-emitter voltage V_{CE}, the input conditions being held constant. Thus the output resistance r_{ce} is evaluated with *constant base current* I_B. r_{ce} is the result of I_C being affected by variations in the base width W, as described in the previous section when the output characteristics in Fig. 5.9(c) were discussed. Thus we can write

$$r_{ce}^{-1} = \frac{\partial I_C}{\partial V_{CE}} = \frac{\partial I_C}{\partial W}\frac{\partial W}{\partial V_{CE}}$$

Both $\partial I_C/\partial W$ and $\partial W/\partial V_{CE}$ are to be evaluated with I_B held constant: the value of $\partial I_C/\partial W$ depends on the relative contributions of recombination and emitter inefficiency to the base current.

In an integrated circuit, transistors are normally made in such a way that recombination in the base can be neglected. In this case, a constant value of base current implies a constant rate of injection of carriers *from* the base into the emitter, as discussed in section 5.5. This in turn implies a constant value of V_{BE}. Then by differentiating eqn. (5.2) for I_C with respect to W with constant V_{BE} we discover that

$$r_{ce}^{-1} = -\frac{I_C}{W}\frac{\partial W}{\partial V_{CE}} \quad \text{(neglecting recombination)} \tag{5.12}$$

The value of $\partial W/\partial V_{CE}$ is negative, due to the *widening* of the depletion layer at the collector end of the base as V_{CB} increases in magnitude. This is base width modulation, described earlier, and the resulting effect on the transistor characteristics is often called the EARLY EFFECT.

Let the width of this depletion layer in the base be W_n. Since the emitter-base depletion layer is much narrower[2] than W_n, the sum of W and W_n must nearly be equal to the physical base width, i.e. the distance between the

[2]The base-emitter junction is in forward bias, which reduces the depletion-layer width substantially.

junctions. Because this distance is constant, $(W + W_n)$ must be constant so that

$$\frac{\partial W}{\partial V_{CE}} = -\frac{\partial W_n}{\partial V_{CE}}$$

Since V_{BE} changes very little in forward bias over the range of base currents normally used, the main contribution to $\partial W/\partial V_{CE}$ comes from the change in collector-base voltage.

The following expression for W_n can be found with the aid of eqn. (3.17):

$$W_n = \left(\frac{2\varepsilon(V_o - V_{CB})}{eN_D[1 + (N_D/N_A)]} \right)^{1/2}$$

Here N_D and N_A refer to the doping concentrations in the base and the collector respectively, and V_o is the built-in voltage at the junction. r_{ce} can be evaluated by differentiating the above equation. Although we shall not quote the resulting equation here, remember that V_{CB} is negative if you wish to evaluate the result.

The computer model used in the program SPICE allows either the use of the above equations to evaluate r_{ce}, or the very simple approximation:

$$r_{ce} = \frac{V_A}{I_C}$$

where the *Early voltage* V_A defined here must usually be specified by the user. This is a reasonably good approximation to real transistor characteristics, a fact which may be anticipated since in eqn. (5.12) neither the base width W nor $\partial W/\partial V_{CE}$ would be expected to change very much with the collector current I_C.

Figure 5.11 shows how the Early voltage may be found from the common-emitter output current characteristics, all of which intersect on the voltage axis at V_A. The reason for this is that their slope r_{ce}^{-1} is proportional to I_C. The diagram exaggerates the slopes for clarity; in reality the value of V_A is often around 50 V or more, giving typical values of r_{ce} upwards of 50 kΩ when the collector current is 1 mA.

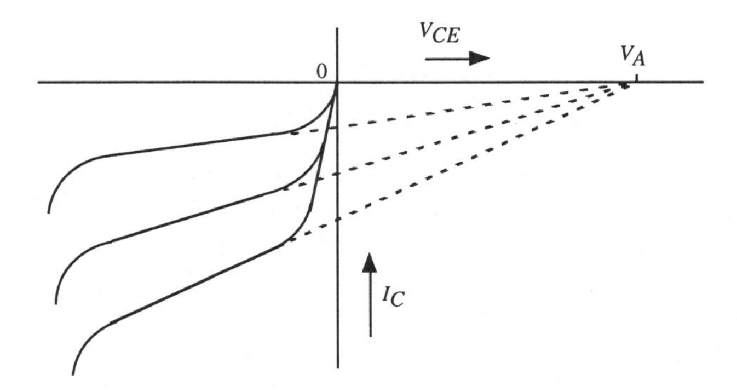

Fig. 5.11 Common-emitter output characteristics, showing the Early voltage V_A. The rapid rise in collector current at high negative collector voltage discussed in Panel 5.1 is also shown.

The size of the output resistance r_{ce} depends on the doping in the base and collector, as seen from the equation for W_n above. r_{ce} can be controlled by the transistor designer to some extent by keeping the *collector* doping N_A low. This ensures that the depletion region lies mostly in the collector[3], minimising both the base width and its variation with voltage, and increasing r_{ce}. As r_{ce} shunts the current generator $g_{fe}v_{be}$, this is a desirable outcome.

Another effect of base width modulation is that very large collector currents may flow at large base-collector voltages. This breakdown phenomenon, called punch-through, is explained in Panel 5.1.

The capacitive components in the small-signal equivalent circuit

If the signals applied to the input terminals of the transistor vary rapidly, the circuit of Fig. 5.10(a) is not an adequate model. The capacitors shown in Fig. 5.10(b) must be added to the low-frequency equivalent circuit. They are closely related to the capacitors used in modelling the junction diode, introduced in Chapter 3.

[3]See section 3. 7

PANEL 5.1

Transistor Breakdown Mechanisms

If base-width modulation were followed to larger and larger base-collector voltages, the depletion layer width would grow, until it became comparable to the physical separation between the collector and emitter junctions. At some particular voltage, the depletion layer would fill the base and this is called "punch-through". At this point, the effective base width W would become zero, whereupon eqn. (5.2) predicts an infinite collector current. This is an example of a means by which failure of the transistor may occur when the currents in it rises above the design values, causing a catastrophic temperature rise.

In practice, a transistor's output characteristic often shows a rapid increase in collector current at values of V_{CB} beyond some critical value, as in Fig. 5.11. This may result from punch-through or from an alternative mechanism called a *current avalanche process*, which is driven by the high electric field present in the depletion region between collector and base during the application of a reverse voltage. A current avalanche arises when collisions of fast electrons with phonons (see Panel 1.3) create sufficient extra electron-hole pairs. The transistor need not be destroyed by these processes, as long as the current is limited by the external circuit, to a value which does not dangerously raise the semiconductor's temperature.

In addition, the small resistance $r_{bb'}$ shown in this circuit is discussed here, since its presence affects the performance of the transistor more at high frequencies than at low frequencies.

Capacitance of the base-emitter junction

First consider the base-emitter junction. It can be treated rather like a diode, so it has a depletion layer capacitance and a diffusion capacitance. When the junction is forward biased, as it is in the active region of the transistor's characteristic, the diffusion capacitance dominates. This capacitance is associated primarily with the charge Q_B stored in the base (eqn. (5.4a)), for the charge injected into the emitter is small by comparison due to the low base doping relative to the emitter. By analogy with the diode, the small-signal diffusion capacitance C_d is proportional to the current I_E through the junction because the charge Q_B is proportional to the current. However, the base charge is usually expressed in terms of I_C rather than I_E. By analogy with the diode,

$$C_d = \frac{e\tau_f I_C}{kT} \tag{5.13}$$

where $\tau_f = W^2/2D_h$ is the *transit time* of minority carriers in the forward direction through the base region, whose width is W. A derivation of eqn. (5.13) is given below. The magnitude of τ_f in real transistors is covered later.

The depletion-layer, or junction, capacitance C_j associated with the base-emitter junction is calculated in an identical way to the diode's. In forward bias it is smaller than the diffusion capacitance C_d, but because the emitter is very heavily doped, thus narrowing the depletion region, it can make a small but noticeable addition to the diffusion capacitance. Thus the total base-emitter capacitance C_{be} is the sum: $C_{be} = C_d + C_j$.

It is primarily the capacitor C_{be}, typically around 1 pF in an integrated transistor, which in conjunction with r_{be} ultimately limits the highest frequency at which the transistor exhibits current gain. For as the frequency of a constant input current i_b rises, the small-signal base-emitter voltage v_{be} in Fig. 5.10(b) falls with the decreasing impedance of the capacitor C_{be}. Consequently, there is a fall in the small-signal collector current generator $g_m v_{be}$. In other words, the short-circuit current gain i_c/i_b falls with frequency, as shown in the log-log plot in Fig. 5.12.

At the frequency f_T the small-signal current gain i_c/i_b falls to unity. f_T is called the GAIN-BANDWIDTH PRODUCT, or sometimes the CUT-OFF frequency. Using the circuit in Fig. 5.10(b), this frequency can be found as follows.

The current i_c may be measured (or calculated) as the current through a short-circuit between the collector and the emitter, so that for completeness we should note that the small collector-base capacitance C_{cb} which is shown

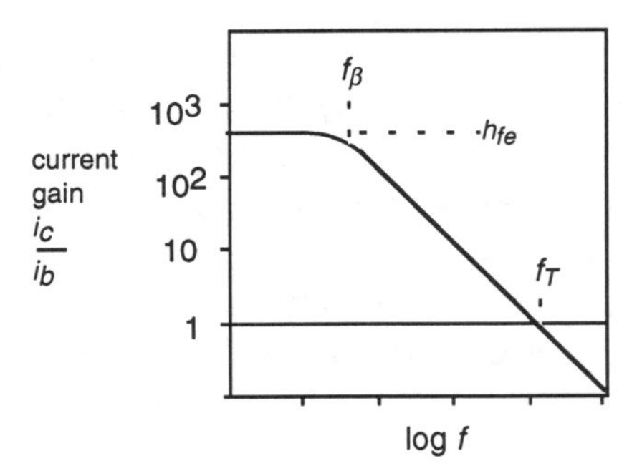

Fig. 5.12 Frequency-dependence of the small-signal current gain in a bipolar transistor.

in the equivalent circuit in Fig. 5.10(b) also appears in parallel with C_{be}. The origin of C_{cb} is explained later.

Since at a sufficiently high frequency f nearly all of the input current i_b flows through the total input capacitance $(C_{be} + C_{cb})$, the voltage v_{be} is given with good accuracy by

$$v_{be} = \frac{i_b}{2\pi f(C_{be} + C_{cb})}$$

Hence

$$\frac{i_c}{i_b} = \frac{g_m v_{be}}{i_b} = \frac{g_m}{2\pi f(C_{be} + C_{cb})}$$

Setting $i_c/i_b = 1$ at the frequency f_T and rearranging gives the result

$$f_T = \frac{g_m}{2\pi(C_{be} + C_{cb})}$$

This expression for f_T can be put in terms of the value of h_{fe} at low frequency, using the relation $r_{be}g_m = h_{fe}$ given in eqn. (5.11). Hence

$$f_T = \frac{h_{fe}}{2\pi r_{be}(C_{be} + C_{cb})} \tag{5.14}$$

Since $C_{cb} \ll C_{be}$, a frequently used approximation to the gain-bandwidth product is

$$f_T \cong \frac{h_{fe}}{2\pi r_{be} C_{be}} \tag{5.14a}$$

Now the frequency

$$f_\beta = \frac{f_T}{h_{fe}} = \frac{1}{2\pi r_{be}(C_{be} + C_{cb})}$$

is the frequency of the point in Fig. 5.12 at which i_c/i_b has the value $h_{fe}/\sqrt{2}$, known as the −3 dB point, or the half-power point. At this frequency, r_{be} and $(C_{be} + C_{bc})$ have equal impedances. Well below the frequency f_β, the current gain i_c/i_b becomes virtually independent of frequency and equal to h_{fe}, since the proportion of the base current which flows into the capacitor C_{be} in the equivalent circuit becomes very small.

The time-constant $r_{be}C_{be}$ of the base-emitter circuit is often written as τ_B, which is also called the *effective lifetime of minority carriers in the base*, for reasons now to be explained. For simplicity, in what follows, we shall ignore the small contributions made by the junction capacitances of both the collector-base and base-emitter junctions.

The significance of τ_B is best seen by first noting that $r_{be} = (\partial V_{EB}/\partial I_B)$ and that, by neglecting C_j, we can take

$$C_{be} \cong (\partial Q_B/\partial V_{EB}) .$$

Thus

$$\begin{aligned}\tau_B = r_{be}C_{be} &= \frac{\partial V_{EB}}{\partial I_B}\frac{\partial Q_B}{\partial V_{EB}} \\ &= \frac{\partial Q_B}{\partial I_B} \\ &\cong \frac{Q_B}{I_B}\end{aligned}$$

when we assume that Q_B/I_B is independent of I_B.

This result, rearranged as $I_B = Q_B/\tau_B$, may be compared with the expression for the *recombination* current I''_B in the base

$$I''_B = \frac{Q_B}{\tau_h} \tag{5.4}$$

where τ_h is the lifetime of holes in the base. Thus, in the absence of any contribution other than recombination to the base current (an unusual situation), we would expect $\tau_B = \tau_h$. In practice, $\tau_B < \tau_h$, since $I_B > I''_B$.

Transistor parameters expressed in terms of the transit time in the base

We now show how τ_B is related in the more general case to transistor parameters, and how C_d (the major part of C_{be}) may be derived analytically. Note particularly how the base charge Q_B plays a central role.

Using eqns. (5.3) and (5.4), the base current at low frequency is given by

$$I_B = \frac{Q_B}{\tau_h} + \frac{AeD_e n_p}{L_e}\left[\exp\frac{V_{EB}}{kT} - 1\right] \tag{5.15}$$

Q_B is proportional to the area under the graph of p_n in the base, as in Fig. 5.4 or Fig. 5.6. Thus

$$Q_B = \frac{WAep_n}{2}\left[\exp\frac{V_{EB}}{kT} - 1\right] \tag{5.16}$$

The expression for I_B above can therefore be rewritten in terms of Q_B:

$$I_B = \frac{Q_B}{\tau_h} + \frac{2D_n}{L_e W}\left(\frac{n_p}{p_n}\right)Q_B \tag{5.17}$$

which we shall rewrite as

$$I_B = \frac{Q_B}{\tau_B} \tag{5.17a}$$

where τ_B, the 'effective lifetime' of holes in the base, is defined by eqn. (5.17a). Comparison with eqn. (5.17) shows that τ_B nearly equals the actual lifetime τ_h of holes if the second term in eqn. (5.17) is small, i.e. if the base current is dominantly due to recombination in the base. Otherwise, τ_B must be smaller than the true lifetime, and its value depends upon all of the parameters in the second term in eqn. (5.17).

Now consider the collector current, which is given by eqn. (5.2):

$$I_C = \frac{D_h eA p_n}{W} \exp \frac{V_{EB}}{kT} \tag{5.2}$$

Comparing this with eqn. (5.16), it should be clear that I_C and Q_B are nearly exactly proportional to one another, since the exponential in eqn. (5.16) is very much greater than unity when the transistor is biased in the active region.

Thus

$$I_C \cong Q_B \frac{2D_h}{W^2} = \frac{Q_B}{\tau_f} \tag{5.18}$$

Here $\tau_f = W^2/2D_h$ is the *transit time* across the base in the forward direction. This is defined in exactly the way that the diode transit time was defined in Chapter 3.

It is interesting to note that eqns. (5.17a) and (5.18) imply a relationship between the large-signal current gain h_{FE} and the time constants defined above:

$$h_{FE} = \frac{I_C}{I_B} = \frac{\tau_B}{\tau_f} \; .$$

Since we know that in real devices h_{FE} varies somewhat with collector current, we can expect τ_B and τ_f to vary too. However, we will ignore this variation here and assume that, when modelling real transistors, a suitably averaged value for each quantity can be used.

Now to change I_C or I_B, Q_B must also be changed. The minority carriers flow in or out via the emitter or collector, but the majority carriers, which must also flow to keep the base region neutral, must enter or leave via the base contact. Thus, at any instant when Q_B is changing

$$I_B = \frac{Q_B}{\tau_B} + \frac{dQ_B}{dt} = \frac{Q_B}{\tau_B} + \left(\frac{dQ_B}{dV_{EB}}\right)\frac{dV_{EB}}{dt}$$

i.e. there is a capacitance $C_d = dQ_B/dV_{EB}$ affecting the rate of change of input voltage V_{EB}. Now

$$r_{be}^{-1} = \frac{dI_B}{dV_{EB}} = \frac{1}{\tau_B}\frac{dQ_B}{dV_{EB}} = \frac{C_d}{\tau_B}$$

so that $\tau_B = r_{be}C_d$. Thus, τ_B is the time-constant of these base-emitter components. Since $C_d = \partial Q_B/\partial V_{EB}$ we use eqn. (5.16) again to find that

$$C_d \cong \frac{eQ_B}{kT} \tag{5.19}$$

Note that C_d, like Q_B, is proportional to the collector current (see eqn. (5.18)), which leads to the result quoted in eqn. (5.13).

It has already been assumed that the junction capacitances are small compared to C_d, as is approximately the case in practice. Then using the expression given in eqn. (5.14a) for f_T we find

$$f_T \cong \frac{h_{fe}}{2\pi\tau_B} = \frac{1}{2\pi\tau_f}$$

This gives a slight overestimate for f_T, but is normally a useful approximation.

Typical magnitudes of C_{be} and the gain-bandwidth product

For a base width $W \cong 1$ μm (typical for a bipolar transistor in an integrated circuit) and taking $D_h = 0.025\mu_h = 10^{-3}$ m²/s, we find that

$$\tau_f = \frac{W^2}{2D_h} = 0.5 \text{ ns} .$$

In practice, τ_f is often smaller still because of a graded doping profile in the base of a real transistor (see Fig. 5.19). This creates an in-built electric field which accelerates the injected carriers across the base. Thus real transistors are faster than the values here would suggest.

Taking $\tau_f = 0.5$ ns, eqn. (5.13) gives $C_{be} \cong C_d = e\tau_f I_C/kT = 10$ pF when $I_C = 0.5$ mA. Then $f_T = 1/2\pi\tau_f = 320$ MHz. In practice, 600 MHz is quite feasible for a silicon transistor with the aid of the built-in electric field.

Capacitance of the base-collector junction

Because the C-B junction is reverse-biased when the transistor is in the active region of operation, it is the junction capacitance (depletion-layer capacitance) which normally dominates C_{cb}. Thus for a junction area A,

$$C_{bc} = \frac{\varepsilon_{\mathrm{Si}} A}{(W_n + W_p)}$$

where W_n and W_p are the depletion layer widths on the *n*-type and *p*-type sides respectively.

Since the collector doping concentration is normally smaller than the base doping, the depletion layer in the collector is wider, and controls this capacitance. Typical values might be in the range of 0.2–1 pF.

Although it is not primarily C_{bc} that determines the frequency at which the short-circuit current gain falls to unity, it often plays a dominant role in limiting the gain obtainable at the highest usable frequencies when the transistor is used in a circuit. For example, in a circuit designed to give voltage gain, the gain at high frequency may become limited by negative current feedback to the base through the collector-base capacitance C_{bc}.

Base spreading resistance $r_{bb'}$

So far we have ignored this component which is often small enough to be neglected. The base spreading resistance $r_{bb'}$ is simply the ohmic resistance of the very thin base, between the *active region* and the *contact*. Because the base is very thin, $r_{bb'}$ can be a few tens (or even hundreds) of ohms, and may limit the rate at which the base-emitter capacitance C_{be} can be charged. $r_{bb'}$ also contributes significant thermal noise at the input of an amplifier, the point where noise is least wanted. For these reasons, it is important that $r_{bb'}$ be kept as small as possible by doping the base as heavily as possible without reducing h_{fe} by too much.

Note that the voltage V_{EB} which appears in all the equations given above should really be written as $V_{EB'}$, where B' is the internal node at the junction of $r_{bb'}$ and r_{be} in Fig. 5.10(b). It is easy to correct those equations if we note that, in the full equivalent circuit, $V_{EB'} = V_{EB} - I_B r_{bb'}$. Hence, by replacing V_{EB} in each of the above equations by $(V_{EB} - I_B r_{bb'})$, they can be readily rewritten to take account into account $r_{bb'}$.

5.11 The charge control model of a bipolar transistor during switching

Let us now examine the effect of the charge stored in the base on the performance of a transistor in a simple switching circuit.

In the circuit of Fig. 5.13, we shall assume that the transistor is switched between on and off states by the input voltage V_{in}, which is switched instantaneously between a high value V_1 (normally a few volts) and a low value $-V_2$ (either zero or less than zero). Figure 5.15(a) shows the waveform of V_{in}. We shall assume that both V_1 and V_2 are very much larger in magnitude than the base-emitter voltage in the transistor's 'on' state. Note that we have

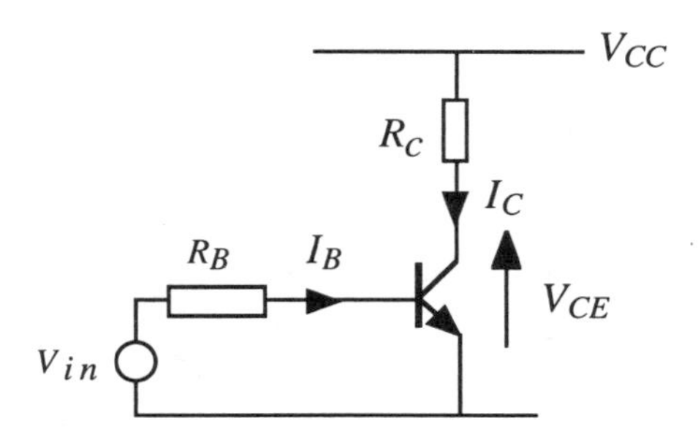

Fig. 5.13 Circuit for investigating the switching time of a bipolar transistor.

chosen to use an *npn* transistor here, so that the base and collector voltages have positive values when the transistor is on.

We expect to find that a finite, measurable time is needed to establish or remove the excess base charge Q_B which is present when when the transistor is fully 'on', and which is illustrated again in Fig. 5.14. The magnitude of Q_B at any instant during switching is related to the base current flowing at that instant.

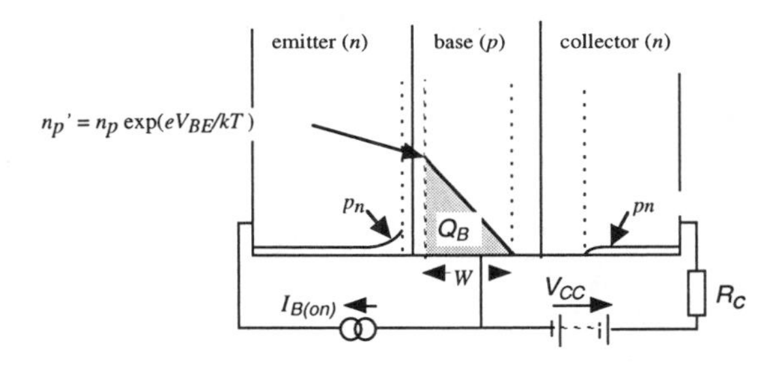

Fig. 5.14 The distribution of minority charge in the base of the transistor, when connected as shown.

Consider first the turn-on process, beginning at time $t = 0$ in Fig. 5.15. Assume that the transistor has been switched off for a very long time, so that V_{BE} begins at the value $-V_2$. On raising the input voltage above zero, a base current flows whose value at any instant is just

$$I_B = (V_1 - V_{BE})/R_B \tag{5.20}$$

The base-emitter voltage V_{BE} rises first from $-V_2$ to zero, in a rather short time. During this time the base-emitter and base-collector junction capacitances C_{je} and C_{jc} are discharged by the base current. The time taken is given approximately by

$$t_d \cong \frac{Q_j}{I_B}$$

where Q_j is the charge stored in the two junction capacitances. This delay is small compared to the rest of the switching time to be discussed next, and may often be ignored. During this time, the base current decreases as V_{BE} rises towards zero, so that there is a short initial transient in the base current waveform, shown in Fig. 5.15(b), before it settles to be nearly constant when V_{BE} becomes positive.

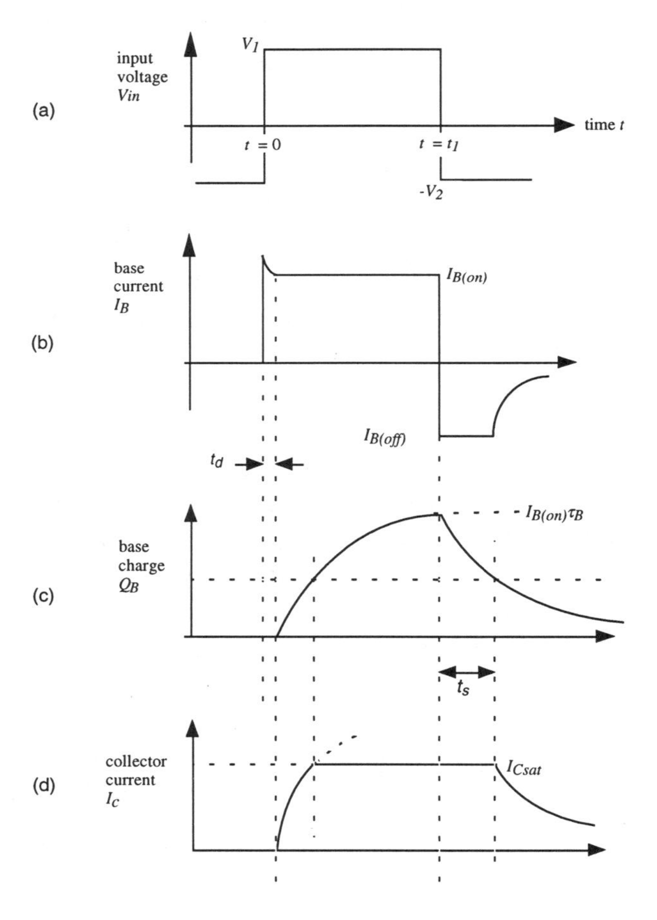

Fig. 5.15 Switching waveforms in the circuit of Fig. 5.13. Waveforms (a) to (d) are described in the text.

Once the base voltage has reached zero, the longer process of injecting the charge Q_B into the base from the emitter begins. During injection, the base current I_B is given by eqn. (5.20) and is approximately V_1/R_B (since we have assumed that $V_1 \gg 0.7$ V). So I_B is nearly constant, as shown in Fig. 5.15(b), after the initial transient during the period t_d. For convenience, we shall call its value $I_{B(\text{on})}$.

Equation (5.17a) relating base current I_B to base charge Q_B is no longer valid, since it was derived for static currents and voltages. In the present dynamic situation, I_B also contributes to the rate of change of the charge Q_B with time. Hence eqn. (5.17a) must be modified by adding dQ_B/dt. Thus

$$I_B = \frac{Q_B}{\tau_B} + \frac{dQ_B}{dt} \tag{5.21}$$

While the base current is held constant at the value $I_{B(\text{on})}$, the solution of this differential equation for Q_B is an exponential one:

$$Q_B(t) = I_{B(\text{on})}\tau_B\left[1 - \exp\left(-\frac{t - t_d}{\tau_B}\right)\right] \tag{5.22}$$

In constructing this solution, we have used two boundary conditions: (i) Q_B is zero initially, at $t = t_d$, (ii) ultimately, when $t \to \infty$, then $dQ_B/dt \to 0$, and hence Q_B tends to the value $I_{B(\text{on})}\tau_B$. Thus Q_B rises as illustrated in Fig. 5.15(c), with the time constant τ_B — the "effective lifetime" which was defined in eqn. (5.15). The collector current rises in proportion to Q_B, as will shortly be shown.

Now we already know that $\tau_B = h_{FE}\tau_f$, the low frequency current gain multiplied by the transit time across the base. We can see that the total time taken for the base charge to rise is very much longer than τ_f, as it is a few times τ_B. The transit time is roughly the time taken for the distribution of minority carriers (holes) across the base to settle into the characteristic triangular shape, illustrated in Fig. 5.14, immediately following the injection of a small additional number of carriers at the emitter junction. Therefore, we can assume that during the injection process, while n_p' is rising, the distribution of minority carriers is very close to the triangular one shown in Fig. 5.14. Because the distribution of carriers is changed so little by the injection process, all the equations relating base charge with currents and voltages are still valid, with the one exception already mentioned, of eqn. (5.17a). The slow rate of injection also means that when using eqns. (5.21) and (5.22), it is valid to employ the same value of τ_B which appears in eqn. (5.16), and which is used in static calculations.

This conclusion also allows us to calculate the collector current from the gradient of the minority carrier distribution, which as before leads to the result $I_C = Q_B/\tau_f$. This is why the collector current follows the exponential rise in Q_B, as we show in Fig. 5.15(d), until the transistor enters the saturation region of operation. That occurs when V_{CE} reaches its lowest value V_{CEsat} of approximately 0.1–0.2 V. The current is then limited by the external circuit (shown in Fig. 5.13) at the value

$$I_{Csat} = \frac{V_{cc} - V_{CEsat}}{R_C} \cong \frac{V_{cc}}{R_C}$$

Otherwise, I_C would continue to rise along the dashed curve in Fig. 5.15(d).

It is important to note that the rate at which I_C rises is controlled not by τ_f, the transit time across the base, but by the effective base lifetime τ_B. As long as the transistor does not enter the saturation region of operation, the delay in turning it fully on is just a little more than $2\tau_B$. Indeed, the 10%–90% risetime is readily shown using eqn. (5.22) to be $2.2\tau_B$.

However, the transistor can be made to enter saturation much sooner than this by increasing the base current $I_{B(\text{on})}$, for example by reducing the value of the resistor R_B in Fig. 5.13. Then the collector current will stop rising when the base charge reaches $I_{Csat}\tau_f$, which is the value given by eqn. (5.18) at the onset of saturation. The base charge Q_B nevertheless continues rising to the value $I_{B(\text{on})}\tau_B$, limited only by the base current.

So if the base current is raised to speed up turn-on, the final base charge, which must be removed during turn-off, is inevitably increased. We shall see later how most of this extra charge can be avoided by using a *Schottky transistor*, but first let us explore the consequences of not doing this.

Figure 5.16 shows the final distribution of minority carrier charge in the base of the saturated transistor, which has risen well above the minimum necessary for saturation to occur (as explained in section 5.9). Since the

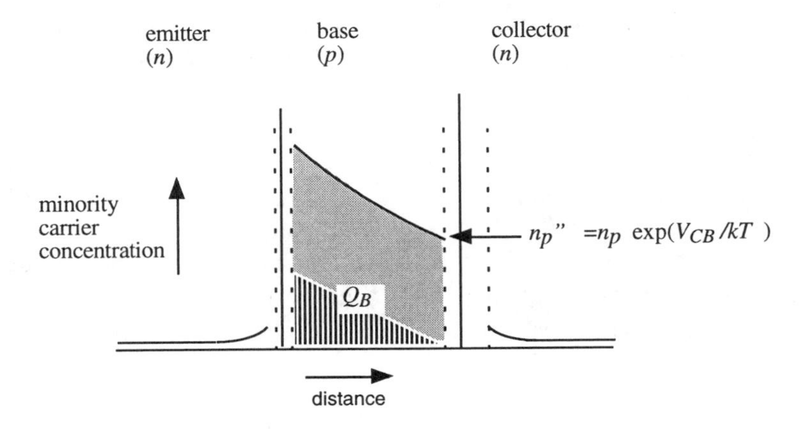

Fig. 5.16 The shaded area is proportional to the charge Q_B stored in the base of a saturated *npn* transistor . The darker shading shows the charge distribution in the unsaturated transistor when carrying the same collector current.

base region is electrically neutral, there is of course a corresponding excess of holes which has been sucked in to compensate the charge on the electrons. We now consider the problem of extracting this charge.

The turn-off transient

Following the course of the waveforms in Fig. 5.15(a), we shall assume that the base charge is fully established by the time t_1 at which the input falls to a negative value of $-V_2$ again. The first feature to note now is that until the base becomes substantially discharged, neither V_{BE} nor V_{BC} can enter reverse bias and must remain positive for a time. The resulting base current is *negative*. Let us call it $-I_{B(\text{off})}$.

Assuming that $|V_2| \gg 0.7$ V, the base current is now

$$-I_{B(\text{off})} = \frac{(V_{\text{in}} - V_{BE})}{R_B} \cong -\frac{V_2}{R_B}$$

The charge in the base begins to decrease at a rate which we can again calculate with the aid of eqn. (5.21), the differential equation for Q_B, using the new value of I_B:

$$\frac{dQ_B}{dt} = -\frac{Q_B}{\tau_B} - I_{B(\text{off})} \tag{5.23}$$

Thus the rate falls as Q_B falls. Note that turn-off can be speeded by making $I_{B(\text{off})}$ large enough. Up until time t_1, the base charge was $Q_B = I_{B(\text{on})}\tau_B$, so that the *initial* rate at which Q_B falls is just equal to

$$\frac{dQ_B}{dt} = -I_{B(\text{on})} - I_{B(\text{off})}$$

The expression for $Q_B(t)$ which results from solving eqn. (5.23), subject to the condition that at $t = t_1$, the initial charge $Q_B(t_1) = I_{B(\text{on})}\tau_B$, is

$$Q_B = I_{B(\text{on})}\tau_B \exp\left(-\frac{(t-t_1)}{\tau_B}\right) - I_{B(\text{off})}(t-t_1) \tag{5.24}$$

This equation, valid only when $t > t_1$, is also sketched in Fig. 5.15(c).

The charge Q_B falls towards the value $I_{Csat}\tau_f$, the value which is the minimum needed for the transistor to be saturated and is represented by the darker shaded area in Fig. 5.16. Until the charge reaches that value, the collector current remains at its saturated value I_{Csat}. The delay before I_C starts to fall from this value can be quite long, since the amount of excess charge is so large and the base current which removes it is normally fairly small, being limited by circuit voltages and external resistances. This delay, which we shall call t_s, is found by setting $Q_B = I_{Csat}\tau_f$ in eqn. (5.24). By rearranging, we find the following transcendental equation:

$$t_s = \tau_B \ln\left(\frac{I_{B(\text{on})}\tau_B}{I_{Csat}\tau_f + I_{B(\text{off})}t_s}\right)$$

This delay before the collector current begins to fall reminds us of the storage time in a diode, which was discussed in section 3.16, and is indeed due to a similar storage mechanism.

The expression for t_s clearly shows that t_s reduces, albeit slowly, with increasing $I_{B(\text{off})}$. To estimate the reduction in the delay attainable in a practical circuit by making $I_{B(\text{off})}$ non-zero, let us assume that $I_{B(\text{off})}$ is made to be equal to $I_{B(\text{on})}$. The cases $I_{B(\text{off})} = 0$ and $I_{B(\text{off})} = I_{B(\text{on})}$ are compared below. For the comparison, $I_{B(\text{on})}$ has been chosen to be twice the minimum required to ensure saturation, i.e. $I_{B(\text{on})} = 2I_{Csat}/h_{FE}$.

(i) When $I_{B(\text{off})} = 0$, the delay becomes equal to $\tau_B \ln(I_{B(\text{on})} h_{FE}/I_{Csat}) = 0.7\tau_B$
(ii) When $I_{B(\text{off})} = I_{B(\text{on})}$, the expression for t_s decreases to

$$\tau_B \ln\left[\tau_B/(0.5\,\tau_f + t_s)\right].$$

When we assume that $t_f << \tau_B$, which is normally true, this result gives $t_s = 0.56\tau_B$.

Because the above analysis assumes that τ_B is a constant, independent of the instantaneous value of the charge in the base, it gives only an approximation of the delays and the measured shapes of the current waveforms of I_B and I_C. However, the behaviour of real transistors can be modelled better if numerical solutions are used with variable τ_B.

The delay t_s is almost entirely avoided in a *Schottky transistor* by incorporating across the collector-base junction a Schottky diode having a low forward voltage drop (about 0.25 V). Figure 5.17 shows the circuit. The diode turns on only when the collector-base junction is forward biased by this amount, and diverts most of the excess base current from the base-emitter junction where it would normally contribute to an increased Q_B. Since the Schottky diode's switching time is due only to its junction

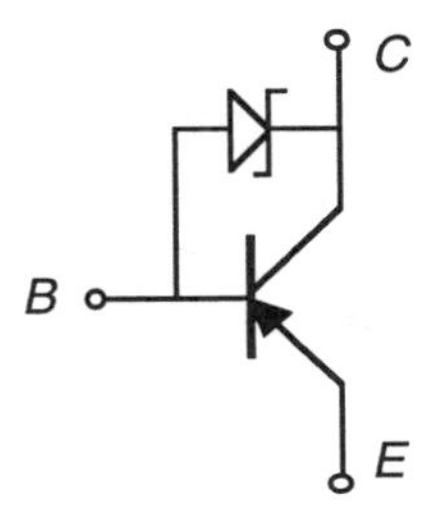

Fig. 5.17 The Schottky transistor consists of a bipolar transistor with a Schottky diode having a low 'on' voltage connected between base and collector.

capacitance, the switching time of the transistor is not limited by the need to discharge the diode. Coupled with the reduced base charge, the result is a much faster turn-off than can be obtained otherwise. Nevertheless, the same principles as we have used here can be applied to the task of modelling the switching time of a Schottky transistor.

5.12 Computer models for bipolar transistors

To construct a mathematical model which can be used in modelling circuits containing many components, a sign convention is required. The usual rules for semiconductor devices are:

(i) all currents *into* a device are positive;
(ii) forward bias voltages are positive, reverse voltages negative, as elsewhere in this book;
(iii) transistor parameters such as g_{fe}, h_{FE}, r_{be} are all positive quantities, but diode reverse saturation currents may have either sign, depending on their forward direction relative to the terminal currents.

The small-signal model cannot, of course, be used to predict d.c. currents from the d.c. voltages (or vice-versa). For computerized circuit simulation, a model is needed which can relate, for example, the collector current to the voltages V_{EB} and V_{CB}, whatever their values. One such equation has already been given in eqn. (5.7), and is repeated here

$$I_C = -I_E\left(\frac{h_{FE}}{1+h_{FE}}\right) + I_{CBO}\left(\exp\frac{eV_{BC}}{kT} - 1\right) \tag{5.7}$$

Fortunately, this already employs the usual sign conventions.

The emitter current I_E can be expressed in a similar way. If the roles of emitter and collector are interchanged, a *reverse* current gain h_{RE} and a saturation current I_{EBO} can be defined, analogous to h_{FE} and I_{CBO} respectively. Then

$$I_E = -I_C\left(\frac{h_{RE}}{1+h_{RE}}\right) + I_{EBO}\left(\exp\frac{eV_{BE}}{kT} - 1\right) \tag{5.25}$$

Since the sign convention states that currents *into* the device are positive, the base current is just

$$I_B = -I_C - I_E \tag{5.26}$$

Equations (5.7) and (5.25) are readily solved for the junction bias voltages V_{BE} and V_{BC} when the terminal currents I_E and I_C are known (and vice-versa, given a little more effort). To specify the transistor, the four parameters I_{CBO}, I_{EBO}, h_{FE}, and h_{RE} are required. In practice, only three of these are needed, since it can be shown that the four are interrelated by the equation

$$I_{CBO}\left(\frac{h_{RE}}{1+h_{RE}}\right) = I_{EBO}\left(\frac{h_{FE}}{1+h_{FE}}\right)$$

The symbol I_S is often used to represent the value of the expressions on either side of this equation, and is the parameter used for the transistor model incorporated into the SPICE simulation program, together with h_{FE} and h_{RE}.

Equations (5.7), (5.25) and (5.26) constitute the full Ebers-Moll model for the large-signal behaviour of the *pnp* bipolar transistor*. Its circuit representation was given in Fig. 5.8, and is shown again in Fig. 5.18(a). The same three equations can be used to model the properties of both *pnp* and *npn* transistors, provided that the saturation currents I_{CBO} and I_{EBO} are made positive in *pnp* transistors and negative in *npn* transistors.

The model which is nevertheless most widely used is a development of the Ebers-Moll by Gummel and Poon, the circuit for which is shown in Fig. 5.18(b). The two circuits can be shown to be exactly equivalent, although the Gummel-Poon model was developed to cater for transistors with a non-uniform doping within each region, and is more flexible in this respect.

As they stand, the Ebers-Moll equations (5.7) and (5.25) do not allow for the effects of base-width modulation. To incorporate that, the parameter I_S must be made to depend upon V_{CB} and V_{EB}. The simplest method is to

*Sometimes the Ebers-Moll equations are quoted in a slightly different but entirely equivalent form, which is more convenient when the terminal currents are to be calculated from the voltages. The interested reader is recommended to refer to more advanced books for details.

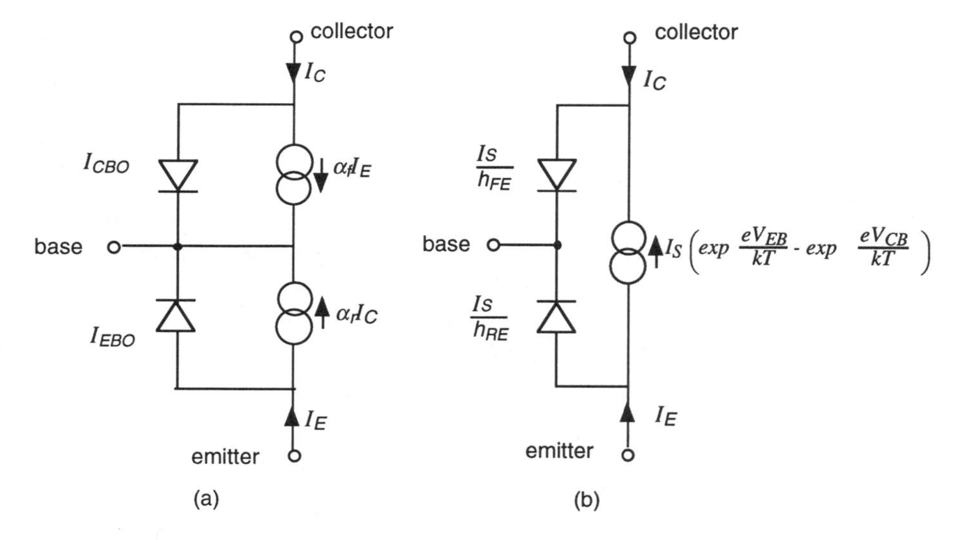

Fig. 5.18 (a) A circuit representation of eqns. (5.7) and (5.25) (b) A circuit representation of the alternative Gummel-Poon model.

introduce two Early Voltages, V_{AF} for forward operation and V_{AR} for reverse operation. I_S may then be replaced with the expression

$$I_S\left(1-\frac{V_{BC}}{V_{AF}}-\frac{V_{BE}}{V_{AR}}\right)$$

For transient analysis, the values of the capacitances C_{be} and C_{bc} at zero junction bias are normally specified by the user, together with the built-in voltages of the two junctions and the exponent m of eqn. (3.20) which are used in that equation to account for the dependence of the junction capacitances on voltage.

For a.c. analysis, once the d.c. operating voltages and currents have been calculated, the small-signal model as in Fig. 5.10(b) is often used. Note that the transconductance g_{fe} need not be provided by the user of a simulation program, since it can be calculated from the expression eI_C/kT. Likewise, the common-emitter input resistance can also be calculated automatically using eqn. (5.11). The small-signal capacitance values are specified as for transient analysis. The energy gap E_g of the semiconductor may also be required by the simulation program to enable it to correct for changes in mobility and carrier concentrations resulting from temperature changes.

Models used for real diffused transistors are necessarily more sophisticated to allow for non-uniform doping, as described in section 5.13. A comprehensive treatment of transistor modelling is beyond the scope of this book, which is aimed at promoting understanding rather than full rigour. For a full description of numerical models, one of the many excellent books on modelling in the reading list should be sought.

5.13 Doping profiles and their effects on transistor performance

Figure 5.19 shows a typical doping density profile in a transistor made by the double diffusion method described in Fig. 5.1. Also shown is a roughly equivalent profile using abrupt junctions as assumed in this chapter. The relative doping levels in emitter, base and collector are normally chosen to optimize as many as possible of the performance criteria given in Table 5.1. The principles of optimization are applicable to both of the profiles in Fig. 5.19. In each case, the performance factor shown on the left of the table is enhanced by using the relative doping levels indicated on the right. Where requirements conflict, a compromise must be effected. Panel 5.2 explains how further improvements have been made possible in the 1990s.

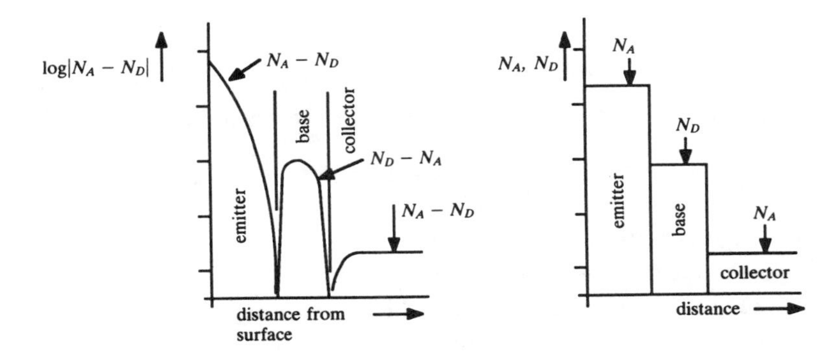

Fig. 5.19 Doping profile of a diffused transistor (left) and an approximate equivalent abrupt doping profile (right).

Table 5.1

Requirement	Desired doping levels in: emitter	base	collector
(1) Low injection into emitter (high emitter efficiency)	high	≤ 1/100 of emitter doping	
(2) Low recombination in base		low	
(3) Low series resistance between base contact and operative base region		high	
(4) Low base width modulation		high	low
(5) Low I_{CBO}		high	high
(6) High collector breakdown voltage (V_{CB})		low	low
(7) Low base-collector capacitance*		low	low
(8) Low series collector resistance			high

*See eqn. (3.19).

PANEL 5.2

Recent improvements to the silicon bipolar transistor

To improve the performance of a transistor, we would like to be able to

(i) reduce the collector-base capacitance to improve high-frequency gain,
(ii) reduce the base spreading resistance,
(iii) keep the value of current gain h_{fe} high,
(iv) keep the output resistance r_{ce} high (i.e. a high Early voltage) by reducing base width modulation.

(*Continued*)

PANEL 5.2 (*Continued*)

These aims are normally realised as far as possible by juggling with the relative doping levels of base, emitter and collector until a suitable compromise is reached. For example, raising the base doping would reduce the base spreading resistance $r_{bb'}$ but lower the current gain.

Recently, it has become possible to raise the base doping and yet achieve sufficient current gain h_{fe}, by causing the emitter to have a *larger energy gap* than either the base or the collector. By this means, injection of carriers from the base back into the emitter is made harder because they have an extra barrier to climb. This is achieved by introducing some germanium into the silicon in the base and collector regions* making a so-called Si-Ge heterojunction transistor. The Si-Ge semiconductor of which the base is made has a lower value of energy gap E_g than that of the pure silicon in the emitter. Because this increases h_{FE} well beyond the value needed in a circuit, the requirement for very low base doping is relaxed.

A higher base doping concentration results in a low $r_{bb'}$ and a high Early voltage. Higher base doping also reduces the depletion-layer width on the base side of the C-B junction, thus reducing base width variations with V_{CB}.

In order to prevent a corresponding increase in the junction capacitance between base and collector, the collector doping is adjusted to ensure that the total depletion-layer width is maintained or even increased. In addition, by increasing the germanium content towards the collector end of the Si-Ge base, an internal electric field is created and the transit time is improved.

In this way, transistors with useful gain at frequencies up to 75 GHz have been made and integrated circuits using this technology are becoming available.

*Special crystal growth methods are needed to prevent the inevitable mechanical stress in this composite structure from creating catastrophic internal defects.

A good compromise is achieved as follows:

(a) Dope the emitter to the limit of solubility of the impurity.
(b) Dope the base as heavily as the desired h_{fe} allows, to satisfy (3), (4) and (5) above.
(c) Dope the collector as little as is consistent with (5) above, so satisfying requirements (4), (6) and (7).
(d) Provide increased collector doping a few tenths of a μm below the collector junction itself to satisfy (8).

Summary of terminology for the bipolar junction transistor

Active region – a region of the characteristics in which the base-emitter junction is forward-biased, while the collector-base junction is reverse-biased.

Saturation region – a region of the characteristics in which *both* junctions are forward-biased.

Common-emitter, or common-base – a connection in which the emitter (or base) is a common terminal for input and output voltages.

Early effect, or base-width modulation – the change in collector current with collector-base voltage.

Ebers-Moll model – a large-signal model for the bipolar transistor.

Cut-off frequency, or gain-bandwidth product – frequency at which the small-signal current gain would fall to unity.

Schottky transistor – a bipolar transistor with a Schottky diode connected between base and collector.

PROBLEMS

5.1 Sketch the construction of an *npn* transistor. Which region is made particularly thin? Is it made thin to achieve (a) low base width modulation (b) high base recombination current (c) high h_{FE} (d) high output resistance?

5.2 State whether the emitter-base and collector-base junctions are forward-biased or reverse-biased in normal amplifying operation. At the emitter-base junction do most of the carriers flow (a) from emitter to base (b) from base to emitter (c) equally in both directions?

5.3 In which regions of an *npn* transistor is the main electron current a drift current, and in which regions a diffusion current?

5.4 Explain the physical mechanisms giving rise to the base current, and show that the principal contributions are proportional to the collector current.

5.5 Is a bipolar transistor saturated when (a) $|V_{CE}| > |V_{BE}|$, (b) $|V_{CB}| > |V_{CE}|$, (c) $V_{CB} > 0$, (d) $|V_{BE}| > |V_{CE}|$, or (e) $V_{BE} > 0$?

5.6 Variation of which voltage or current causes base width modulation?
(a) I_C (b) V_{CE} (c) V_{BE} (d) V_{CB} (e) I_E .
Base width modulation is the origin of which of the following?
(a) r_{be} (b) g_{fe} (c) r_{ce} (d) h_{fe}.

5.7 Give the preferred order of increasing doping levels in the emitter, base and collector. Which of the following assists in obtaining (i) high h_{fe} (ii) low base width modulation?
(a) high base doping relative to collector
(b) high base doping relative to emitter
(c) high collector doping relative to base
(d) high emitter doping relative to base.

The next five problems refer to the same transistor:

5.8 An *npn* bipolar silicon transistor has a base doping of 10^{23} m^{-3}, an effective base width of 1.5 μm and a junction area of 0.1 mm^2. Make a dimensioned sketch of the electron concentration in the base region when the collector current is 5 mA and the collector junction is reverse-biased. What base-emitter voltage will sustain this collector current?

5.9 The effective recombination time of electrons in the base of the above transistor is 4×10^{-8} s. What is the recombination current in the base?

5.10 What *minority* carrier density in the emitter will ensure that the hole current across the base-emitter junction does not exceed the base recombination current? Assume a hole recombination length of 2 μm in the emitter. How is this density achieved in practice?

5.11 Explain, with the aid of a diagram showing carrier densities, why the common-emitter output characteristics of a junction transistor have a finite slope. An *npn* silicon bipolar transistor has a base doping of 10^{23}, an effective base width of 1.5 μm, and collector doping which is very much higher than in the base. Initially V_{CB} is set to 5 V. What reverse bias across the collector-base junction will increase the collector current by 10%, when I_B is held constant? Consider the two cases, when I_B is dominated by (a) recombination in the base and (b) emitter inefficiency, i.e. injection of holes into the emitter. You may neglect the built-in voltage of the collector junction. Base width modulation is often modelled as a resistor in the hybrid-π equivalent circuit. Why is this only an approximation? (i.e. why is the slope of the output characteristic not constant)?

5.12 A small forward voltage of 0.59 V is applied across the collector-base junction of the transistor in problem 5.11, while the collector current is held at 5 mA. Sketch the new minority carrier distribution across the base. How is I_B affected by the forward bias on the collector?

5.13 Given a bipolar junction transistor with unlabelled leads, one variable d.c. voltage power supply and one multirange ammeter, how would you determine whether the transistor was *npn* or *pnp*, and which lead was which?

5.14 In a particular bipolar transistor, h_{fe} is 100 at low frequency, and the frequency at which the value of h_{fe} extrapolates to unity is 100 MHz. Estimate the transit time of minority carriers across the base, the diffusion capacitance at a collector current 5 mA and the frequency at which the effective value of h_{fe} is $1/\sqrt{2}$ times the low-frequency value.

5.15 In Fig. P5.1, two ways of connecting an *npn* transistor as a diode are shown. Sketch the minority hole and electron concentrations through the transistor in each case for identical bias voltages. Assume the following: doping $N_D(\text{emitter}) = 4N_A(\text{base}) = 8N_D(\text{collector})$; $D_e = 2D_h$; $W_E = W_B = 0.5\ W_C$; and identical junction areas. Neglect recombination (i.e. assume infinite lifetimes).

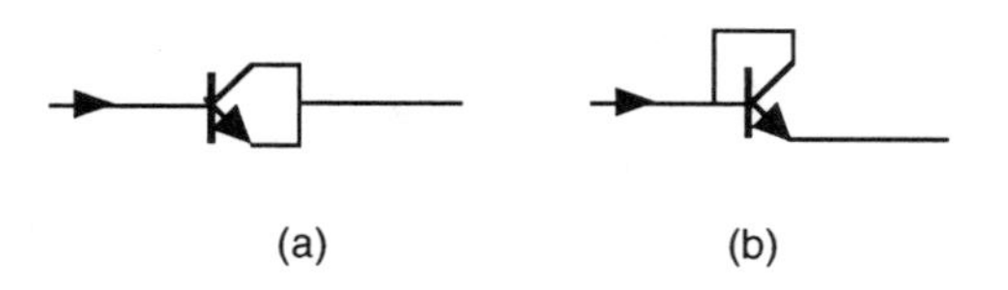

Fig. P5.1

Which connection gives, at the same bias voltage:
(i) the largest current?
(ii) the smallest total excess minority carrier charge?
(iii) the smallest diffusion capacitance?
Give your reasons in each case.

5.16 Explain why the transit time across the base of a bipolar transistor controls the high-frequency current gain. In a particular bipolar transistor h_{fe} is measured to be 30 at a frequency of 20 MHz. Estimate the base transit time and the collector current which gives a base-emitter diffusion capacitance of 20 pF. The base-collector *junction* capacitance can reduce the gain of an amplifier at high-frequency by permitting current feedback to the base from the collector. How should doping levels be adjusted to minimise its value?

Chapter 6

Integrated Device Fabrication

The characteristics of real devices differ somewhat from those of the 'ideal' devices we have discussed in earlier chapters. One prime cause of these differences is the way that devices are normally mass produced, using the 'planar process', which revolutionized electronics by making integrated circuits possible. This process results in a distribution of dopant atoms which differs from those we have assumed for the analysis of device behaviour.

In this chapter we describe the techniques used in fabrication, and show how the distribution of impurities is both controlled and calculated. We shall need to make use of several now familiar ideas, such as thermal energy and diffusion. After outlining steps in a complete fabrication process, the two processes of impurity implantation and diffusion will be discussed in more detail, in order that the reader may appreciate some of the advantages of current technology, and some of the limitations it imposes on device performance.

6.1 A simplified silicon microcircuit fabrication process

As an example of a complete fabrication process, we shall outline a simplified schedule of processes for fabricating low-power NMOS transistors. The description which follows differs in its details from the procedures used for making integrated circuits in industries, but the sequence described below nevertheless illustrates most of the basic elements of a real fabrication process. The latter will contain many more steps than are shown here, even though a manufacturer is as much concerned to simplify production as he is to improve the performance of his products. The main steps are illustrated in Fig. 6.1, which shows a series of cross-sections through part of a thin slice, or WAFER, of silicon. We assume that the slice, 100 mm–200 mm in diameter, has already been cut from a large cylindrical crystal of *n*-type silicon using a diamond saw, and that its surfaces have been carefully polished until they are very flat.

Step I

The 200–300 μm thick wafer of *n*-type silicon is first given a weakly doped *p*-type surface layer about 10 μm thick,* in which the individual components are to be formed. We shall see later that this step is convenient in order that all components in a circuit may be electrically isolated from one another.

Step II

A layer of SiO_2 (silica) is grown on the surface in a furnace at a temperature of about 1100°C using one of the following two reactions:

$$Si + O_2 \rightarrow SiO_2$$

$$Si + 2H_2O \rightarrow SiO_2 + 2H_2$$

This oxide is subsequently used to prevent the introduction of dopants into the silicon below, in places where they are not wanted.

---

*For the chemical and physical methods used for this and other processes see e.g. 'Materials Science' by Anderson *et al.*, Chapman and Hall, 4th edition, 1990.

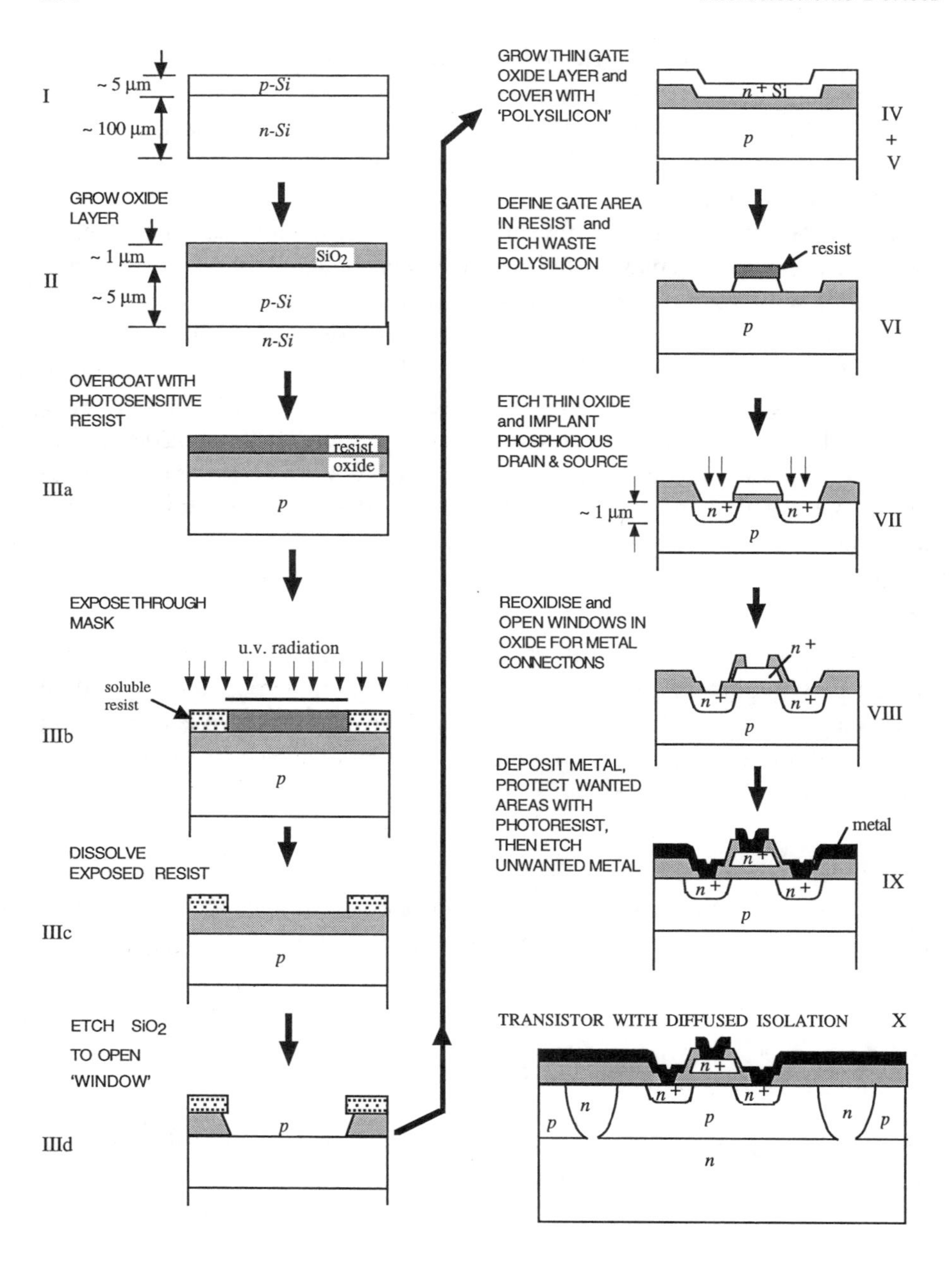

Fig. 6.1 Steps in the manufacture of an *n*-channel MOS transistor.

Step III

'Windows' through to the silicon are opened in the oxide layer by a sequence of steps (Diagrams III(a)–(d)) which are often repeated at later stages in the process.

First, the SiO_2 is covered with a 1 μm layer of photosensitive polymer called a PHOTORESIST. For application, it is dissolved in a solvent which evaporates rapidly. The dried photoresist has the property of becoming more soluble in another organic solvent (called the developer) on exposure to ultra-violet light. Exposure takes place through a mask, an ultra-violet image of which is focussed onto the wafer. The mask prevents exposure to the light in areas where the resist will be required on the surface for the next step in the process. The mask may be simply a photographic emulsion or, more commonly, a metallic film on the surface of a silica glass plate. The mask is appropriately patterned, as illustrated in Fig. 6.2, so that many identical transistors and their associated connections can be fabricated simultaneously.

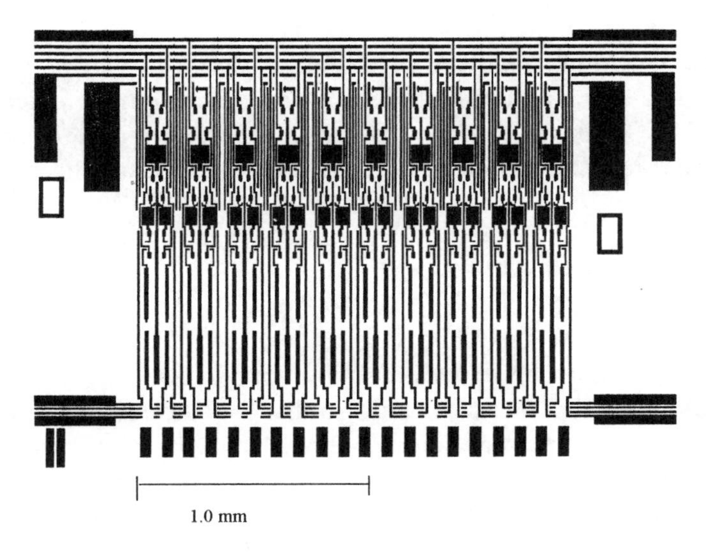

Fig. 6.2 A portion of the pattern of a mask used to shield photoresist from ultra-violet radiation. The area shown represents a small portion of the interconnection pattern for an integrated circuit. *Courtesy of P. Roberts.*

After exposure, the soluble, exposed photoresist is dissolved away (step III(c)) in the developer solvent, leaving the SiO_2 uncovered in the area where the 'window' is required. The latter is then etched through in its turn in step III(d), usually by a process called REACTIVE ION ETCHING. This involves heating the wafer inside a sealed, partially evacuated chamber in which it is exposed to a fluorocarbon-rich mixture of gases through which an electrical discharge passes. The discharge and the high temperature together cause the fluorine and carbon atoms to react with the SiO_2, forming volatile SiF_4 and CO_2. The polymeric mask resists etching by this process, but is afterwards removed in a similar equipment, by using an oxygen-rich gas mixture to oxidise the hydrocarbon photoresist to CO_2 and H_2O.

Step IV

The silicon is now available for processing in the location where the MOSFET is to be formed, while the surrounding area is protected by the SiO_2. The next step is to form the gate oxide, which is made by the same procedure as the oxide in step II. However, it is made much thinner, about 50 nm.

Step V

Now the gate material itself is deposited over the *whole* wafer. The material is a heavily doped *n*-type polycrystalline form of silicon, called POLYSILICON in the business, which is deposited on the wafer at high temperature by the reaction of gaseous tetrachlorosilane with hydrogen:

$$SiCl_4 + 2H_2 \rightarrow Si + HCl$$

Step VI

The gate area is to be defined next by applying photoresist as in steps III(a)–(d), but with a different mask, so that the hard resist is left covering the gate. The unwanted polysilicon is etched away in the surrounding areas by a suitable reactive ion etching process, which preferentially removes Si but leaves the underlying SiO_2 largely untouched, where it is exposed by removing the polysilicon. Usually this process involves the use of a chlorine-bearing gas mixture, which is tailored to selectively etch silicon but not its oxide.

Step VII

On either side of the gate itself lie the regions where the source and drain will be formed, covered only by a thin layer of gate oxide. These regions are now implanted with an *n*-type dopant by the use of a large piece of equipment called an ION IMPLANTER, described in more detail in section 6.3. Inside this, the wafer is exposed, in vacuum, to bombardment by a high-velocity beam of arsenic or phosphorous ions, which penetrate the oxide and embed themselves in the silicon at an average depth of between 0.1 μm and 0.3 μm below the surface. Subsequent treatment of the wafer at high temperatures may cause these dopant ions to end up at a greater depth, since they can diffuse through solid silicon at temperatures approaching 1000°C. At the same time, the dopant atoms can diffuse sideways under the gate oxide itself, so making the source-drain separation slightly smaller than the width of the gate itself, which masks the channel of the transistor from the implantation process.

Resistors may also be made elsewhere on the wafer at this stage, their values being determined partly by the concentration of dopant introduced by the process of implantation, and partly by their dimensions.

Step VIII

The wafer is next prepared for making metal connections to the source, drain and gate. The silicon is first re-oxidized, and photoresist is again used to open windows in the oxide, where the connections are required.

Step IX

Aluminium is next deposited over the whole surface, by placing the wafer in a vacuum chamber, in which a source of pure aluminium is heated to vapourize it. The aluminium vapour condenses on the cool wafer until a thickness of about 1 μm is built up, in the space of less than a minute.

This metal layer is then patterned to interconnect the transistor with other components by using photoresist once more, finally etching away the unwanted metal with a suitable acid. Contact pads for external connections about 100 μm square are formed at this stage.

The final steps (not illustrated) usually include the following:

(a) A protective glassy coating is applied over the whole circuit.
(b) The wafer is sliced into chips, which are then mounted on holders which also support the legs or pins which form the external contacts.
(c) Contact to the pins is made with gold or aluminium wire which is bonded by cold welding both to the aluminium pads on the chip and the pins.
(d) The chip is embedded in plastic, or hermetically sealed under a metal or ceramic cover, if wide temperature excursions are expected in use.

Step X

Component isolation. No conduction is possible between different transistors made in the same *p*-type substrate, provided that the latter is connected to the most negative part of the circuit in which they are used. However, the threshold voltages of the transistors may then differ, since they may have different source-substrate voltages. For this and other reasons, it is useful to be able to isolate different parts of the *p*-type substrate from one another. This can be done by separating the various *p*-type regions by 'walls' of *n*-type impurity as illustrated in the final diagram in Fig. 6.1. These are diffused from the surface right through to the *n*-type wafer below the *p*-type layer grown on the surface. This is normally done *before* fabrication of any components, because the deep diffusion requires a very high temperature.

The total number of steps in the process is large enough for there to be a significant risk of device failure. The YIELD of good devices (or of functional chips containing many devices) is necessarily less than 100 per cent of those fabricated. This is a determining factor in the price of the final product, and the number of process steps is always kept to a minimum, in order to maximize the yield.

6.2 Bipolar transistor fabrication

We shall not describe each and every step of making a bipolar transistor, and shall confine the description to the *npn*-type. This is usually favoured over its *pnp* brother because the higher mobility of electrons permits faster on-off switching, and, correspondingly, amplification to higher frequencies.

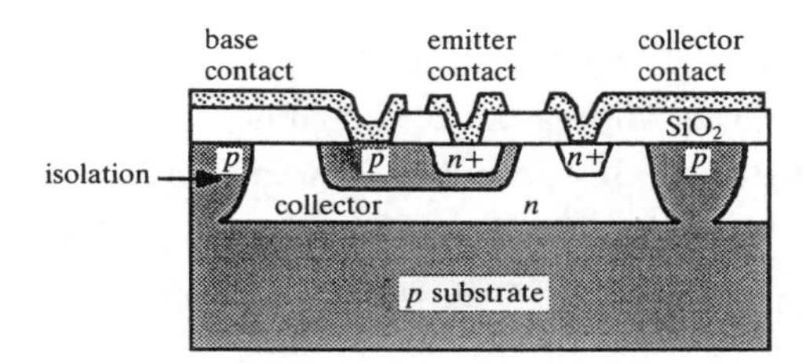

Fig. 6.3 Construction of an *npn* bipolar transistor showing isolation diffusions.

An *npn* transistor is commonly made in an *n*-type surface layer (an epitaxial layer) as indicated in Fig. 6.3. Since this layer forms the collector, it may be necessary first to isolate neighbouring transistors by a *p*-type isolation diffusion which penetrates to the underlying *p*-type wafer.

The base region is then formed by implanting and diffusing e.g. boron to make a junction at a depth of about 1–3 μm. The doping concentration at the surface must not be too high, so that the subsequent phosphorous doping for the emitter region can turn the surface *n*-type. This phosphorous doping is controlled so as to form the emitter-base junction at a depth of about 0.5–2 μm. This ensures that the base in thin enough to give a high current gain. Control of the base width is made easier if the phosphorous concentration can be made to fall more steeply with depth than does the boron concentration. This can be achieved using ion implantation.

A typical doping profile in such a transistor was sketched in Fig. 5.10. At the same time as forming the emitter, phosphorous is also implanted into regions where contacts are to be made to the weakly *n*-type collector layer. This is necessary, in order that no rectifying contact occurs when aluminium is subsequently deposited to connect the transistor to other components (see Panel 3.II).

6.3 Ion implantation

Ion implantation is currently (1997) the principal technique used to introduce *p*-type and *n*-type impurities into silicon. It creates a very thin layer of doped semiconductor close to the surface, usually much less than 1 μm thick. This method is also used to adjust the threshold voltage of MOS transistors by accurately doping the thin channel region itself. If the doping is required to a greater depth, subsequent heating can cause atomic diffusion, so that the impurity atoms may penetrate deeper into the semiconductor. Diffusion will be discussed later, in sections 6.5–6.7.

The dopants commonly implanted in silicon are boron, arsenic and phosphorous. The equipment is designed to produce a beam of fast-travelling ions of the dopant, which is aimed at the wafer so that the ions crash into the surface and bury themselves there.

In an evacuated chamber, labelled A in Fig. 6.4, a gas of the required species is ionized by passing an electric current through it. The ions are allowed to diffuse out of the small chamber containing the electric discharge, into another evacuated chamber. There they are accelerated by a high voltage (between 20 kV and 1 MeV) towards a hollow electrode, through which they pass into a region of high magnetic field, directed at right angles to their path. The magnetic field bends their path so that only those with a specific charge/mass ratio will pass through an exit hole and fall with high velocity on the wafer. A mask deposited on the wafer, made of glassy SiO_2 or just a coating of photoresist, can be used to prevent doping where it is unwanted. The total dose of ions per unit area of the wafer is easily measured,

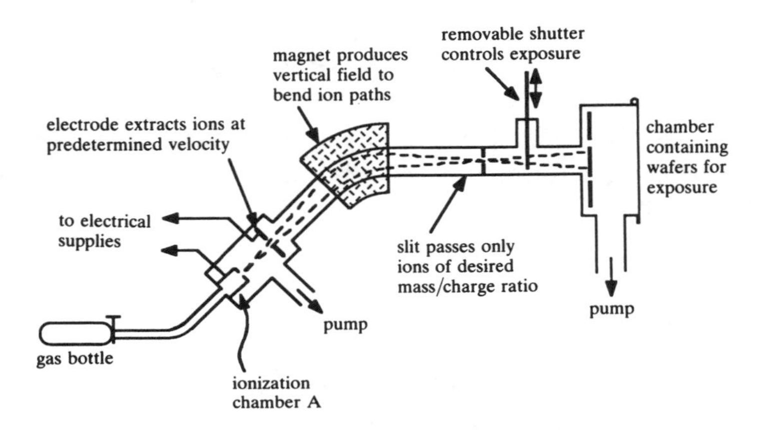

Fig. 6.4 Schematic diagram of ion implantation equipment.

since it is just proportional to the electric current I carried by the ions, and to the total time of exposure, t:

$$N\ (\text{per unit area}) = \frac{It}{eA}$$

where A is the wafer area. The current flowing from the wafer to an earthed electrode is readily measured by means of an ammeter.

Typically, the total beam current might be 1 mA or more, corresponding to a flux of 6×10^{15} singly charged ions per second. Each ion crashes into the silicon with high energy (perhaps 100 keV) and high momentum, and stops only after colliding with many silicon atoms. As more than 2 eV of energy on average is transmitted to each silicon atom, most such atoms will be displaced from their normal positions, and the resulting damage in the crystal would create traps for holes and/or electrons, if the crystal were not repaired.

The damage is repaired by gently 'cooking' or ANNEALING the wafers at a temperature which gives the silicon atoms just enough energy to move back into their correct locations relative to their neighbours. Luckily this can be done at a low enough temperature that the dopant atoms are unable to move very far from their initial positions by diffusion.

The distribution of the dopant concentration with depth is discussed next.

6.4 Implanted doping profile

An incoming ion is deflected by every impact on a silicon atom, so that it follows a very zigzag path. This is illustrated in Fig. 6.5(a). The depth of penetration below the surface, called the RANGE, is shorter than the length of path travelled. Its average value for all ions is controlled by the accelerating voltage used. Naturally, there is a random spread of penetration depths, or 'STRAGGLE'. The random element leads to a distribution of the concentration of stopped ions which is spread about an average range R, as illustrated by the distribution plotted in Fig. 6.5(b) as a solid line on a log-linear graph. The concentration $N(x)$ of ions which penetrate to a depth x

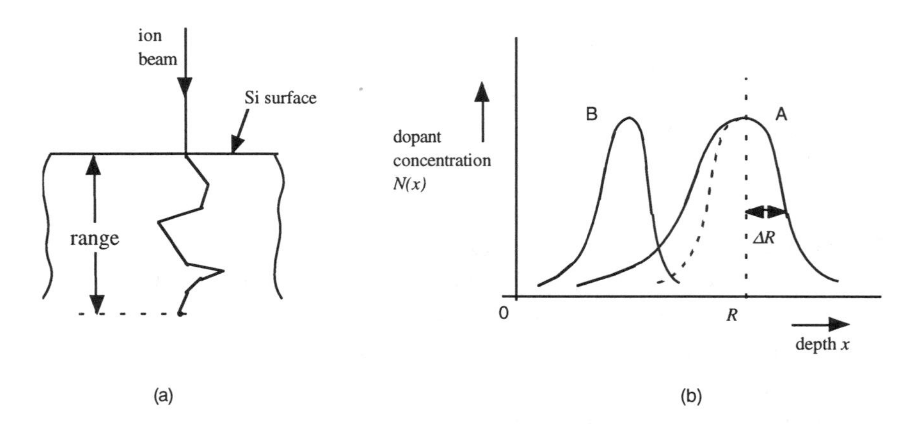

Fig. 6.5 (a) Typical path of an implanted ion (b) resultant depth profile of stopped ions (solid line A) compared with a gaussian approximation to it (dashed line). Curve B is the depth profile for a much lower incident ion energy.

is shown for two ion energies. A simple mathematical approximation which is sometimes used to model the measured data is a symmetrical, *gaussian* distribution, which can be expressed in terms of the mean range R and straggle ΔR as follows:

$$N(x) = N_0 \exp \frac{-(x-R)^2}{2\Delta R^2}$$

This is quite a good approximation for depths greater than R, but it underestimates the concentration $N(x)$ at smaller depths, as shown by the dashed line in the figure.

The parameters R and ΔR are usually determined experimentally; the range R varies from, for example, 0.01 μm at 10–12 keV energy for large ions like arsenic and phosphorous, to about 0.5 μm at 200 keV energy for the small boron ion. The straggle ΔR ranges from about 0.1R to 0.5R, depending on the dopant and the energy used.

To produce the very shallow implant needed for MOSFET threshold control, the dopant is often implanted *through* the gate oxide itself, with an energy chosen so that the peak of the distribution is located at the surface of the silicon.

6.5 Diffusion of impurities

The high temperatures involved in nearly all fabrication steps which follow after the implantation of a dopant species result in redistribution of the dopants. Thus, there is a change in the concentration profile within the device being made. These changes result from diffusion of the impurity atoms within the solid silicon. Diffusion may also be used deliberately to redistribute the dopant, driving it to greater depths below the wafer surface.

Diffusion, whether in gases, liquids or solids, results from the randomizing effects of thermal kinetic energy. Heat always tries to make the concentration of a chemical species uniform everywhere, just as when two gases or liquids are mixed together. In solids, diffusion might appear difficult, if not impossible. But a covalent solid like silicon is an open lattice structure with space between atoms (see Fig. 1.2) through which and into which atoms may move. For example, silicon atoms occasionally break loose from their normal positions, having by chance acquired more than the average thermal energy. They leave behind a vacant site, or VACANCY, so that at any temperature a small but finite proportion of all the silicon sites is vacant. This is illustrated in Fig. 6.6. There is a well-defined concentration of vacancies (just as of holes), which rises with temperature. These vacancies will migrate (slowly) due to continual thermal agitation, i.e. they diffuse through the solid.

If a high concentration of impurity atoms is established at the surface, some will find their way into vacancies present in the surface layer of silicon atoms. If another vacancy in its random motion reaches an adjacent

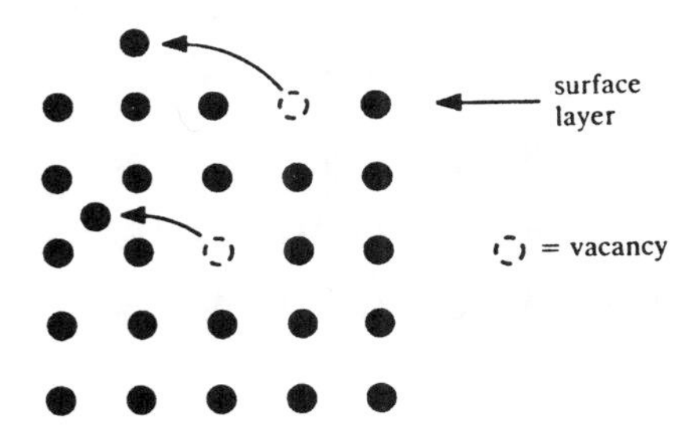

Fig. 6.6 Illustrating vacancy formation and diffusion in a schematic silicon lattice.

site, the impurity atom may then move into it, taking a new position. This happens only when the impurity atom has acquired enough thermal energy to overcome the force holding it in position. So motion does not occur instantly, though the more readily as the temperature is raised. Thus the rate of diffusion of impurities rises rapidly with temperature.

Solid state diffusion in a concentration gradient

To see that the above process leads to a rate of flow proportional to the concentration gradient, consider Fig. 6.7. Assume that the probability P that an impurity atom jumps to a neighbouring vacant site in unit time is dependent on

(a) the probability that the neighbouring site is vacant
(b) the probability that the impurity atom has enough thermal energy

and NOT on

(c) the concentration N of impurity atoms
(d) the direction of the vacant site from the impurity.

In Fig. 6.7 the plane AB lies normal to the concentration gradient, which is in the x direction. The rectangle AB is taken to have unit area, and lies midway between two planes of atoms separated from one another by the distance a. If the concentration of impurities is N at the plane of atoms to

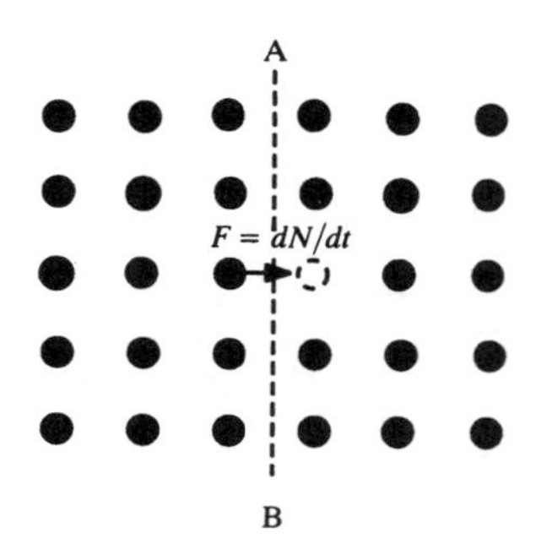

Fig. 6.7 Calculating the flow of impurity atoms in a concentration gradient.

the left of AB, then at the plane to the right it is $N + a(dN/dx)$. The flux of impurity atoms across AB from left to right is then NP, whilst the flux from right to left is $(N + adN/dx)P$.

The net flux F in the x-direction is thus

$$F = P\left(N - \left(N + a\frac{dN}{dx}\right)\right) = -Pa\frac{dN}{dx} \; .$$

As with electrons or holes we define a DIFFUSION COEFFICIENT D by putting

$$F = -D\frac{dN}{dx} \tag{6.1}$$

D is usually measured rather than calculated, and depends on

(a) the nature of the diffusing species
(b) the medium in which diffusion occurs
(c) the temperature.

All of these affect the probability P.

We look next at the temperature dependence of diffusion. This may be bypassed on a first reading, as an understanding of it is not necessary for subsequent discussion.

6.6 Temperature dependence of diffusion in solids

Looking back at the assumptions laid out at the beginning of the previous section, the probability P that an impurity atom will move depends on two energies: the energy E_v needed to create a vacancy on a neighbouring site (about 2 eV), and the energy E_b needed for the impurities to temporarily break three bonds with its neighbours before they are remade with the new neighbours. The latter is also about 2 eV.

The probability of an atom having thermal energy E or more is not given by the Fermi function, but can be shown to be proportional simply to $\exp - E/kT$. Hence we can see that the probability P is proportional to both

$\exp - E_v/kT$ and $\exp - E_b/kT$, i.e.

$$P \propto \exp - \frac{(E_v + E_b)}{kT} .$$

The diffusion coefficient therefore varies in the same with temperature; thus we can put

$$D = D_0 \exp \frac{-E_a}{kT}$$

where the prefactor D_0 is almost independent of temperature and $E_a = (E_b + E_v)$, called the ACTIVATION ENERGY of diffusion, is usually in the range 3.5–4 eV. Note that, at the temperatures of interest, around 1300 K, the diffusion coefficient rises by $2\frac{1}{2}$–3 per cent per Kelvin, a behaviour which is quite different from the diffusion either of gases or of electrons and holes. The diffusion coefficients of the common dopants in silicon are shown in Fig. 6.8 as a function of temperature. The strong dependence on both temperature and dopant has important implications for the sequence of steps in a manufacturing process, for, if doping with

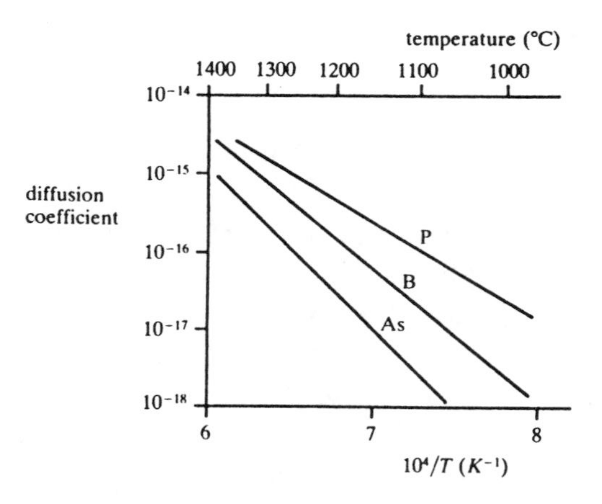

Fig. 6.8 Diffusion coefficients of boron, phosphorous and arsenic in intrinsic silicon as a function of temperature.

a particular dopant species is followed by a process needing a higher temperature, substantial further diffusion will occur.

6.7 Doping profiles after diffusion

Diffusion from the surface of a wafer is determined by the temperature and by the initial distribution of impurity at the surface prior to diffusion.

The resulting depth profile of the impurity concentration may be found mathematically from the solution of a differential equation, using an appropriate boundary condition which is determined by the supply of dopant at the surface. Qualitatively, we can understand the shape of the profile with the aid of the analogy of heat diffusion introduced earlier.

If a thin layer of impurity is placed at the surface and allowed to diffuse into the wafer, the depth distribution develops as in Fig. 6.9.

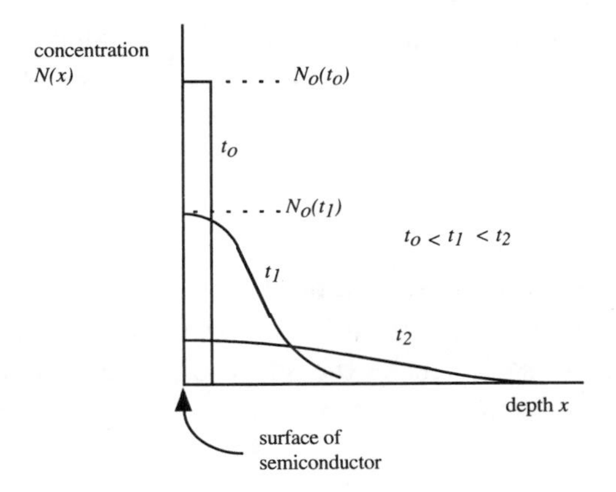

Fig. 6.9 Development with time of the doping profile arising from a fixed quantity of dopant initially at the surface. $t_2 > t_1 > t_0$.

The problem is similar to that of finding the temperature profile in a bar initially at uniform temperature, after a thin slab of hot metal is placed in contact with one end. (No heat is assumed to flow into the air.)

Since the quantity of heat, or impurity, is constant, the area under each curve is the same, and each can be shown to have a GAUSSIAN shape: the concentration $N(x)$ at depth x below the surface is given by

$$N(x) = N_0 \exp -\left(\frac{x^2}{a^2}\right) \tag{6.2}$$

where N_0 and a both depend upon time t. Although at the starting time t_0 this equation is not an exact description, it is close enough for practical purposes. It can be shown that the following solution satisfies these conditions:

$$N(x, t) = \frac{Q}{\sqrt{(\pi Dt)}} \exp -\left(\frac{x^2}{4Dt}\right) \tag{6.3}$$

where Q is the total number of impurity atoms deposited per unit surface area at the outset. Comparison with eqn. (6.2) shows that the surface concentration N_0 at time t falls as $t^{-1/2}$. Note also that, at depths less than about $\sqrt{(Dt)}$ the concentration is little below its value at the surface. At depths greater than twice this, the concentration falls rather steeply.

Junction depth

Figure 6.10 shows the complete impurity profile in an *n*-type wafer into which has been diffused e.g. boron from a fixed source quantity as above. Note that the log–lin scales make the shape of the acceptor doping profile into a parabola, since from eqn. (6.3):

$$\ln \frac{N}{N_0} = -\frac{x^2}{4Dt}$$

As can be seen from the diagram, the depth x_j at which the junction is formed is the solution of the equation

$$N_D = N_A(x,t) = \frac{Q}{\sqrt{(\pi Dt)}} \exp -\frac{x_j^2}{4Dt}$$

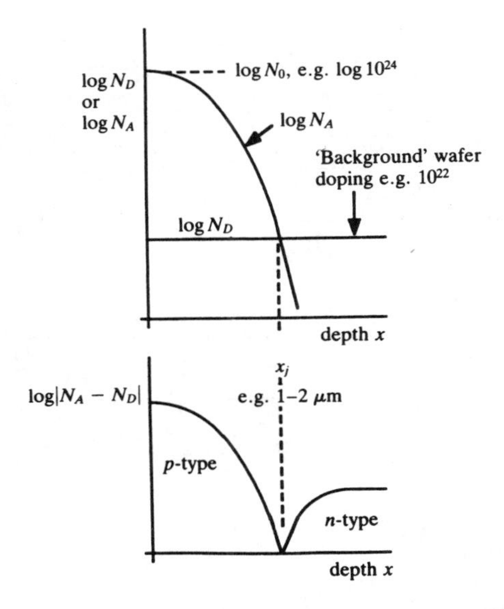

Fig. 6.10 Log-linear plot of a complete doping profile, showing the formation of a junction at depth x_j.

Since the pre-exponential factor depends only weakly on time t, the junction depth x_j rises approximately as $t^{1/2}$. It also rises rapidly with the temperature at which diffusion occurs, since D is strongly temperature-dependent.

6.8 Integrated circuits

The techniques of making integrated circuits follow rather closely the pattern outlined earlier in this chapter for making MOSFETS. There are many other aspects of integrated circuit design and fabrication to which justice cannot be done fully here.

Figure 6.11 illustrates in both plan and cross-section the construction of two types of integrated circuit: one uses bipolar transistors and the other comprises both p-channel and n-channel MOSFETS — a so-called complementary MOS circuit (often abbreviated to CMOS).

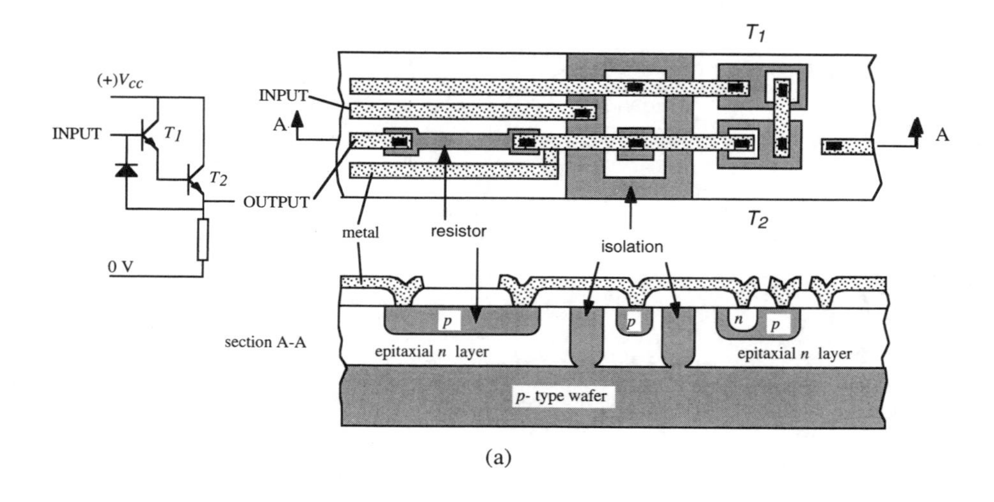

(a)

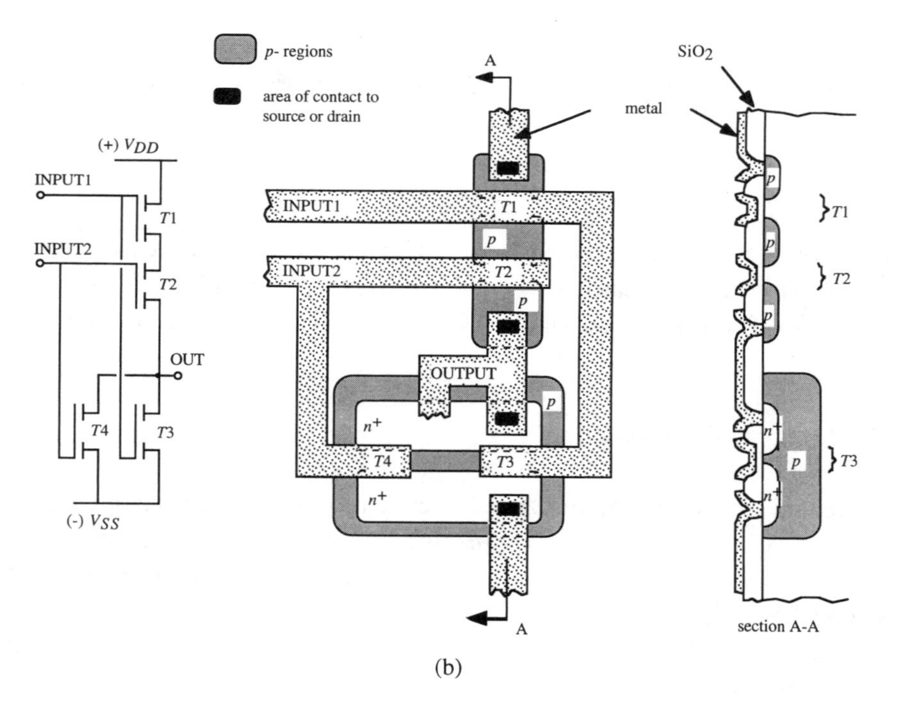

(b)

Fig. 6.11 Two hypothetical integrated circuits in both plan and section (a) a bipolar transistor circuit (b) a complementary MOS circuit.

There are several features to note: the construction of resistors, the means used for isolating components from one another, and the conductor cross-overs. These last can be achieved by making connections via the silicon. Without cross-overs, the circuit complexity would be severely limited; it is common to find metal conductors crossing polysilicon conductors, too, thus providing a third level of interconnection between components.

Capacitors are generally avoided in circuit design, unless very small values will suffice. It is difficult to produce more than about 10^{-3} pF of capacitance per square micrometre of chip area. Some types of semiconductor digital memory, however, make use of the small gate capacitance of the MOSFET structure, and may contain millions of such devices.

Large-scale integrated circuits employing close-spaced components are sensitive to minor disturbances in the fabrication process. Slight misalignments in the relative positions of conductors, contacts and diffusion regions, or the presence of dust particles where insulating SiO_2 is grown, can cause the yield of good devices or circuits to fall below acceptable levels. The greatest care is thus needed in Very-Large-Scale-Integration (VLSI) manufacture, where the sensitivity to defects is greatest.

Nevertheless, it is to be remembered that a wafer of silicon 200 mm in diameter has a surface area sufficient to make more than 1000 chips, each of which measures 5×5 mm and comprises typically 10^6 components. If say, 20 wafers are processed at a time, then a yield of good chips of only 20 per cent of those made produces 4000 functioning chips in each batch. It is this, coupled with the high reliability of such circuits, which gives this technology its major advantage.

The reader who wishes to find out more about the design and manufacture of integrated circuits is advised to consult one of the many specialised books on the subject. He will find that a good grasp of the contents of this book will assist him to understand much in the more advanced texts.

PROBLEMS

6.1 What are the two main functions performed by the layers of SiO_2 formed on a silicon wafer?

6.2 What are the particular properties which make 'photoresist' suitable for its task?

6.3 What is a vacancy? How does it assist in the process of diffusion of impurities into silicon?

6.4 Why are the dimensions of diffused resistors slightly greater than those of the mask used to define their boundaries?

6.5 Describe the sequence of processes needed to fabricate a resistor and its contact in (but isolated from) an *n*-type epitaxial layer.

6.6 The diffusion rates of impurities in silicon increase in the sequence As, B, P, Al. What two impurities could be selected in each of the following cases?

(a) The *p*-well and source-drain diffusions in an *n*-type substrate to form an *n*-channel enhancement FET.

(b) The base and emitter diffusions in a *p*-type substrate to form a *pnp* transistor.

6.7 Why is an n^+ implant made into *n*-type regions of a wafer before an aluminium contact is made to it? Would this be necessary for all contact materials?

6.8 Define the terms 'range' and 'straggle'.
What process must always follow ion implantation, and why?

Further Reading

The following books contain more extensive treatments of diode and transistor behaviour, as well as discussion of various devices not covered in this book.

Bar-Lev A 1993 Semiconductors and Electrical Devices, 3rd edn., Prentice Hall.

Tyagi M S 1991 Introduction to Semiconductor Materials and Devices, Wiley.

Shur M 1990 Physics of Semiconductor Devices, Prentice Hall.

For a clear and comprehensive treatment of MOSFETs, the following is recommended:

Glasser L A, Dobberpuhl D W 1985 VLSI Circuits. Addison Wesley, ch 2.

For a more detailed account of silicon processing and integrated circuits, see:

Morgan D V, Board K 1983 Introduction to Semiconductor Microtechnology, Wiley.

For a more detailed discussion of the computer model SPICE, see:

Massobrio G, Antognetti P 1993 Semiconductor Device Modelling with SPICE, 2nd edn., McGraw Hill.

Answers to Numerical Problems

Chapter 1

1.2 (a) N, P, As, Sb, Bi
(b) B, Al, Ga, In, Tl

1.3 C, Si, Ge, Sn, Pb

1.4 5×10^{23} m^{-3}, zero, 10^{-5}

1.5 (a) 0.032 Ωm, 1.2×10^{18} m^{-3}
(b) 0.013 Ωm, 2×10^{10} m^{-3}
(c) 6.6×10^{-3} Ωm, 1.2×10^{17} m^{-3}

1.6 1.5×10^{22} m^{-3}

1.7 4.6 kΩ

1.8 2×10^{9} m^{-3}

1.10 2.6×10^{12} s^{-1}, 1.3×10^{12} s^{-1}

1.12 *p*-type, 0.18 m^2/Vs

1.13 1.7×10^{23} m^{-3}

Chapter 2

2.8 5×10^{-9}, $(1.0^{-5} \times 10^{-9})$

2.11 1.7×10^{18} (Si), 4.3×10^{20} (Ge)

2.12 about 10^{-19} m^{-3}
2.13 0.16 μm
2.14 0.87 μm, 0.84 μm
2.15 about 4×10^{21} m^{-3}

Chapter 3

3.1 (c)
3.6 (c)
3.7 (b)
3.8 0.25 V, 1.35 V
3.9 0.64 V, 0.25 V
3.10 0.392 pA, 4.56 μA
3.12 0.3 ms
3.13 2.9 μm, 91 nm (Si); 2.1 μm, 66 nm (Ge)
3.14 0.3 V, 39.2 mA, 0.45 A
3.16 (a) $\times 1/\sqrt{2}$
(b) $\times 1/\sqrt{6}$, $\times \sqrt{2/3}$
3.17 1.9 pF, 0.12 μF
3.18 45 ns, 13 μm
3.19 4.6 kV/m, 0.06 V, 10

Chapter 4

4.1 (a)
4.2 (d)
4.3 (b), (c) are correct
4.5 (a)
4.9 about 0.064 pF
4.12 0.31 mS
4.13 0.17 pF, 0.42 mS/V
4.14 3.2 V, 1.4×10^{-3} F/m^2
4.15 3.5
4.19 14 ps, 40 ps

4.21 34 nm, 0.28 mA/V
4.23 2.8×10^{-3} F/m^2, 3.7×10^{16} m^{-2}

Chapter 5

5.1 (c)
5.2 (a)
5.5 (d)
5.6 (d)
5.7 (c)
5.8 0.647 V
5.9 40.2 μA
5.10 $\leq 6 \times 10^7$ m^{-3}
5.11 about 19 V, 4 μA
5.12 between +4 μA and +8 μA
5.13 (a) $V_{CB} = 7.7$ V
(b) $V_{CB} = 10.4$ V
5.14 1.6 ns, 307 pF, 1 MHz
5.16 270 ps, 1.85 mA

Chapter 6

6.6 (a) B, P
(b) As, B or P, Al

Index